SpringerBriefs in Earth Sciences

For further volumes:
http://www.springer.com/series/8897

James P Terry · A Y Annie Lau · Samuel Etienne

Reef-Platform Coral Boulders

Evidence for High-Energy Marine Inundation Events on Tropical Coastlines

 Springer

James P Terry
A Y Annie Lau
Department of Geography
National University of Singapore
Singapore

Samuel Etienne
École Pratique des Hautes Études
    CNRS Prodig
Dinard
France

ISSN 2191-5369   ISSN 2191-5377 (electronic)
ISBN 978-981-4451-32-1   ISBN 978-981-4451-33-8 (eBook)
DOI 10.1007/978-981-4451-33-8
Springer Singapore Heidelberg New York Dordrecht London

Library of Congress Control Number: 2013931646

Printed on acid-free paper

Springer is part of Springer Science+Business Media (www.springer.com)

*He could hear the heavy boom of surf
against the broken shore and see how the
great billows thundered down upon the
naked coast
Then did the knee-joints and courage of
Odysseus fail him, and sadly he questioned
his own brave spirit*

Homer, The Odyssey

Preface

In recent years, the study of high-energy marine inundation (HEMI) events on tropical coastlines has become increasingly popular as researchers try to develop a better understanding of coastal hazards, or more to the point, whether the coastline was inundated by a tsunami or a storm. The topic of boulder deposits has become something of a divisive issue in the coastal hazard community and as such this book could not come at a more crucial time. Furthermore, much of the work is being carried out in tropical regions and a summary of the state of affairs is not only timely, but will serve to create a level playing field which in itself will doubtless serve to dispel several misconceptions.

Public and scientific attention has been focussed on the devastating effects of recent tropical HEMI events, the most noteworthy of which are undoubtedly the 2004 Indian Ocean, 2007 Solomon Islands and 2009 South Pacific tsunamis, the long reach of the 2011 Tōhoku-oki tsunami, and a litany of annual tropical cyclones and typhoons. Many of these have generated reef-platform coral boulder deposits, but to focus on these alone in the absence of an historical context is dangerous. It is often said that the one thing we learn from history is that we learn nothing from history. It is therefore wonderful to see that once the authors get past the introduction and scope they get stuck in straight away to an historical review and changing terminology which provides the key context to the long (over 200 years) history of coral boulder studies. However, the recognition of coral boulders as something special is not just reserved for recent scientific studies and they have long been noted in Traditional Environmental Knowledge and local terminology. The Fijian *Vatu ni Cagi Laba* and the Japanese *tsunami-ishi* are just two examples of such knowledge.

Perhaps one of the most difficult areas for the non-specialist to understand, and even for that matter some of the specialists, is how a boulder gets onto the reef-platform. Yes they are invariably (but not always—see Chap. 4) placed there by some form of HEMI, but to understand the process starts to give the scientist and the risk manager the kind of information they are looking for—was it a storm or a tsunami? There has been considerable debate surrounding the use (and misuse) of various hydrodynamic equations to determine the transport process and it is therefore excellent to see this topic get a thorough review. All researchers embarking on boulder research would do well to consult this work. The authors however are not willing to leave any skeletons in the cupboard and so the ability to provide a chronology for the events that placed the boulders on the reef platform in the first place also gets an airing. This is a vexed topic which is also plagued by the inappropriate use of dating techniques and sample selection. Boulders come and boulders go—indeed the same process that puts a boulder on a reef platform can also take it away, or a different process can do that, so we are always looking at an incomplete picture of events that have occurred over what is often an indeterminate time period. Does this negate the value of boulder research altogether? No. What this book does so well is highlight the potential pitfalls that both the wary and unwary researcher can face while also showing what can only be described at the "best practice" approach to recording and studying these anomalous geomorphological features. One of the great things about doing boulder research correctly is that when the fieldwork is over you have all the data you need—there is no need for samples to be taken back (fortunately) for further analysis. However, to be in such a position it is important to go in knowing what to do and the authors offer a case study from Fiji that illustrates this well.

Finally, the authors make a key point that the efficient and effective study of boulders lends itself to the development of standardised datasets that can be compared and analysed in a way that will undoubtedly significantly advance our understanding of boulder emplacement. I thoroughly recommend this book to anyone undertaking or wanting to undertake coastal boulder research.

Prof. James Goff
Director, Australia-Pacific Tsunami Research Centre
University of New South Wales
Sydney
Australia

Acknowledgments

The authors extend their thanks to the following people for their kind contributions and assistance given freely in various ways, without which the research and preparation of this book would not have been possible.

Dr. Adam Switzer, Earth Observatory of Singapore
Dr. Alan Ziegler, Department of Geography, National University of Singapore
Dr. Bruce Richmond, United States Geological Survey, Santa Cruz, USA
Prof. Gennady Gienko, Department of Geomatics, University of Alaska Anchorage, USA
Mr. Iliesa Qoli, Bouma village, Taveuni Island, Fiji
Mr. Iosefu Golea Soroalau, Lavena village, Taveuni Island, Fiji
Mrs. Lee Li Kheng, Cartographer, Department of Geography, National University of Singapore
Ms. Lee Ying Sin, Earth Observatory of Singapore
Ms. Marie Augeyre, formerly University of French Polynesia, Tahiti
Dr. Mark Buckley, United States Geological Survey, Santa Cruz, USA
Dr. Michael Gregory, The University of the South Pacific, Suva, Fiji
Dr. Raphaël Paris, Centre National de la Recherche Scientifique, Clermont-Ferrand, France
Mr. Romano Reo, Chief Surveyor, Lands and Survey Department, Kiribati
Mr. Tevita Mafi, Lavena village, Taveuni Island, Fiji

In addition, there are many other individuals, colleagues, family members and friends, to whom the authors owe a debt of gratitude. All are thanked for their insightful comments, logistical help and constant support.

Research funding is gratefully acknowledged from the following sources:

The National University of Singapore and Singapore Ministry of Education (grant number FY2012-FRC2-005 and Doctoral Research Scholarship); Hong Kong: Hui Yin Hing Fellowship; French Authorities (Haut-Commissariat de la République) and the Territory of French Polynesia (CPER Rinalpof, 2011–2013); The Ministère des Affaires Etrangères (Fonds Pacifique—French Embassy in Suva, Fiji, 2009); The University of French Polynesia.

Contents

Chapter 1
Coastal Boulders: Introduction and Scope

Abstract Boulders represent a singular class of sediment, encountered in very diverse geomorphic systems. In coastal areas, they may appear as isolated exotic deposits or as a part of larger constructional features. Owing to their size, boulders are less easily reworked than finer sediments and their presence has become increasingly recognised as an important signature for high-energy marine inundation (HEMI) events. They therefore represent a valuable subject for natural hazard studies. This is especially evident along tropical coasts where coral reefs may provide large quantities of boulders that accumulate on the reef flat or farther inland. This volume explores the significance of a particular subset of coastal boulders, namely reef-platform coral boulders. While these geomorphological objects have been scrutinised in the last decade following catastrophic events like the 2004 Indian Ocean tsunami, they actually bear a much longer history in terms of scientific interest stretching back over two centuries.

1.1 Types of Coastal Boulders

Coastal boulders are large clasts of rock that have been detached from bedrock sources and subsequently transported to their depositional sites in coastal settings. Where present, large boulders are a conspicuous element of coastal geomorphology and have consequently been described on many types of shorelines around the world. Coastal boulders may exist as isolated objects, or if occurring in sufficient numbers they may form extensive boulder fields strewn across the intertidal flats. Elsewhere, boulders are part of the fabric of coastal constructional features such as storm ridges or rubble ramparts comprising a mix of coarse sediment fractions. Frohlich et al. (2009, 2011) describe coastal boulders as 'erratics', indicating that they possess lithologies different from the country rock type. Where this is indeed the case, any one of a range of mechanisms may have been responsible for their delivery to coastal locations, for example fluvial or glacial transport, volcanic activity or ice rafting. Shanmugam (2012), however, prefers to use the term 'exotic boulders' underscoring the fact that they are often prominent features of the coastal landscape, even if their lithology is consistent with their immediate environment.

Not all coastal boulders are true erratics. A number of geomorphic processes operating in coastal environments are capable of locally producing large clasts with lithologies that match the surrounding geology. These processes include in situ weathering at the coast (McKenna et al. 2011), exhumation of 'pseudo-boulders' produced by unusual types of diagenesis such as concretion (Fig. 1.1), rockfalls from coastal cliffs or other types of mass movements (Fig. 1.2), or wave erosion at the shoreline (Fig. 1.3). Where coastal boulders are indeed of marine origin, high wave energy is normally required for their detachment and transportation. As such, the presence of boulders has become increasingly recognised as an

Fig. 1.1 The Moeraki Boulders on the Otago coast of South Island, New Zealand (45°20.9′S 170°49.6′E). The stretch of beach where these strikingly spherical boulders are seen is now protected as a scientific reserve. Marine erosion of the Palaeocene-age mudstone cliffs has exhumed these pseudo-boulders, mostly 0.5–2.2 m in diameter, which have accumulated on the beach in clusters. Their formation results from in situ diagenetic alteration (concretion) within the mudstone rather than hydrodynamic erosion of the rock mass, hence their coastal location could lead to their mistaken interpretation as true marine boulders. Photos by J. Terry, January 2008 (*left*), S. Etienne, August 2011 (*right*)

Fig. 1.2 Volcanic boulders at the UNESCO World Heritage Site known as the Giant's Causeway in Northern Ireland. The biggest boulders with coloured lichen cover, although partially rounded, come from the cliff slope as revealed by their singular lithology. They were transported to the shoreline as rockfalls. Photo by S. Etienne, May 2012

Fig. 1.3 Water-worn boulders with smooth faces on the beach at Matei (16°41.3′S 179°52.6′W) on the northern tip of Taveuni Island in Fiji. The boulders are the eroded remains of Quaternary basaltic lava flows that project from the coastline as rocky promontories. Photo by J. Terry, July 2010

important signature for high-energy marine inundation (HEMI) events, principally intense storms (e.g. Morton et al. 2006; Richmond and Morton 2007) and tsunamis (e.g. Dawson 1994; Paris et al. 2009). Accordingly, with careful interpretation, coastal boulders can be effective tools for identifying ancient HEMI events, while detailed investigation of their size, location and age may provide clues as to the magnitude and timing of the original events that produced them.

1.2 Association with High-Energy Marine Inundation Events

According to 2003 data, 2.385 billion people live within 100 km of the coast, which represents 41% of the global population (Martínez et al. 2007). For a variety of reasons, a steadily increasing proportion of the world's population is either choosing to live, or is finding itself living, at the coast. This in itself is not a problem except when communities are exposed to multiple natural hazards that can potentially cause significant impacts on coastlines, such as erosion, river floods, saltwater intrusion and marine surges due to tsunamis or tropical storms (Goff and Terry 2012). For illustration, in recent times it is the coastal zone that has witnessed the horrors of the December 2004 Indian Ocean tsunami (IOT) and the March 2011 Tōhoku (Japan) Tsunami. Yet, in spite of the occurrence of such devastating recent events, it is still not always easy to appreciate the risks faced by many low-lying coastal areas, either because no historical records exist of past inundations from which we might learn valuable lessons, or because societies remain ignorant of a particular area's geological past (Goff and Terry 2012). Thus, as with all types of natural hazards, a primary goal in coastal hazard research (and coastal management) is to improve our understanding of both the magnitude and frequency of past HEMI events, in order to be better prepared for the future. Although HEMI events have clearly occurred throughout pre-historical times, for many vulnerable coastlines limited information is available on how often or how large those events were. Even where historical records do exist, a major hurdle that impedes vulnerability assessment is their relatively short duration.

Within this context, sedimentological analysis of marine inundation deposits remains among the most valuable approaches for teasing out the palaeo-record of past HEMI events (Terry and Etienne 2011). During an intense storm or tsunami, an assortment of sedimentary material is scoured from the seabed and intertidal area and deposited at the coast or farther inland. Perhaps surprisingly, it is notable how most studies aiming to understand coastal impacts by investigating sediments have concentrated mainly on fine-grained deposits such as sand sheets or buried sand layers. In contrast, HEMI-emplaced coastal boulders are generally under-represented in previous research (Paris et al. 2011). For instance, Scheffers et al. (2009) commented that while photographs showing large boulders dislodged by tsunamis do appear in some publications, discussion of these coarse-grained deposits is mostly missing. "This leads to the false presumption that modern tsunamis have moved only fine sediments and that boulders moved by these tsunamis are not preserved or recognised" (Scheffers et al. 2009, p. 553). This might be unexpected, considering that easily entrained fine-grained sediments are more likely than boulders to be washed by waves into mangrove swamps, coastal lagoons and vegetated back-beach environments, where they become intercalated with existing sediments, mixed up by bioturbation, or altered by pedogenesis, all of which adds complexity to the interpretation of HEMI events through sedimentology.

In light of the above, scientific interest in the origin, transport and emplacement of large coastal boulders has been growing apace over recent years. This is especially so on tropical coastlines, where blocks of reef rock torn off the seaward edges of coral reefs by tsunami or storm waves and thrown up onto exposed reef platforms have obvious advantages in terms of identification and sampling, compared to more elusive layers of sands and gravels buried in mangroves, obscured by shoreline vegetation or submerged in coastal lagoons (Terry and Etienne 2011). Moreover, the carbonate framework of coral boulders offers potential for age-dating, although establishing links between the ages of carbonate boulders and the time of their emplacement on coasts also presents its own set of challenges. Another benefit is the greater longevity that boulder-sized clasts sometimes exhibit over finer sediments. Paris et al. (2010), for example, found that boulder accumulations brought by the 2004 IOT were the only surficial evidence left after two years at Lhok Nga Bay (southwest of Banda Aceh in northern Sumatra, Indonesia), in spite of the enormous volume of fine material carried on land by this exceptionally powerful tsunami. This is because sediment accretion was controlled by coastal processes that were influenced by post-event meteorological conditions, such as longshore drift currents, and as a result much of the fine-grained sediments were remobilised during a relatively rapid phase of beach recovery (Wassmer et al. 2007).

1.3 Rationale for this Book

What has been mentioned in the preceding section provides the overall rationale for this book. At present it is clear that there is still much we need to learn about the nature of HEMI events, especially those that predate the historical record. Closely connected with this, it is also evident that the examination of coastal boulders of marine origin offers one important line of investigation for geomorphologists who wish to interpret the nature and characteristics of past extreme events on coasts, for which neither local knowledge nor documentary evidence exists. Not all coastlines are, of course, equally vulnerable to marine inundation and neither do all places have the same potential for the production and long-term preservation of bouldery deposits. This is one reason why tropical coastlines are chosen as the focus here. Given the right conditions, tropical coastlines are fringed by living coral reefs, which are a source of coarse clastic carbonate sediments (i.e. coral boulders) that can be quarried by high-energy wave action (Fig. 1.4). Coral reefs likewise provide suitable platforms at sea level where this erosional debris can be deposited. Henceforth, this particular subset of coastal carbonate boulders will be referred to as 'reef-platform coral boulders' (RPCBs).

In summary, the primary goal of this volume is to present a treatise on the significance of RPCBs (Figs. 1.4 and 1.5) for understanding both modern and prehistorical (Holocene) HEMI events on tropical coastlines. This is an exciting new field that intersects tropical coastal geomorphology and natural hazards science. Although there has been a groundswell of interest in large carbonate boulders on tropical coasts over the last decade, it is not widely appreciated that such features were observed and recorded during the explorations of Matthew Flinders on the Great Barrier Reef back in the early 1800s. Our intention is to demonstrate how various characteristics of RPCBs yield valuable evidence about the swells, storms and tsunamis that emplaced them over centennial timescales. No comprehensive review has yet been published, so it is anticipated that this work will fulfil the

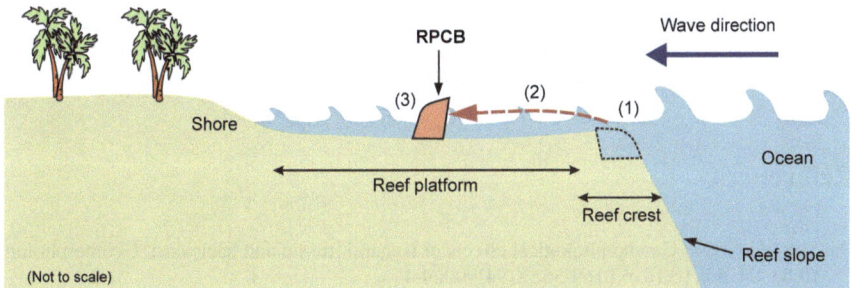

Fig. 1.4 One possible mode of formation for a reef-platform coral boulder (RPCB). Living coral reefs are the usual source of these boulders on tropical coastlines with adjacent fringing reefs. Most RPCBs are produced when part of the reef crest or reef slope is detached (1), transported (2) and then deposited (3) on the reef platform by high-energy wave action

Fig. 1.5 A large reef-platform coral boulder (RPCB) that was deposited on the coast at Lhok Nga (5°29′N 95°15′E) in north western Sumatra, Indonesia, by the 26 December 2004 Indian Ocean tsunami (IOT). Photo by S. Etienne, August 2006

need for a definitive reference on coral boulder research, which details the earliest observations, changing terminology, progress in sedimentology and growing relevance for hazard studies on tropical coastlines. Although the greater portion is given over to description of RPCBs and their value for interpreting HEMI events, other boulder lithologies seen on tropical coastlines (e.g. volcanic boulders) are also considered where published work illuminates cross-cutting aspects of boulder research that are more broadly relevant, such as wave transport modelling. Numerous examples are cited from around the world, although there is an emphasis on Asia, Australia and the Pacific basin as part of the up-to-date review of the existing literature. A case study of original findings from recent investigations by the authors on the island of Taveuni in Fiji is finally presented to highlight current methods and emerging themes.

Readers will note that the main arguments presented in the text are supported wherever possible with illustration: line figures, maps, photographs, tables, graphs, charts and other forms of imagery. This design is deliberate, so that this book is not only of interest to Geomorphologists and Earth Scientists, but also makes the information contained within accessible to non-specialists who are more generally concerned with extreme events, natural hazards, tsunamis, tropical cyclones, coral reefs and the changing nature of tropical coastlines over time.

References

Dawson AG (1994) Geomorphological effects of tsunami run-up and backwash. Geomorphology 10:83–94. doi:10.1016/0169-555X(94)90009-4

Frohlich C, Hornbach MJ, Taylor FW (2011) Megablocks. In: Hopley D (ed) Encyclopedia of modern coral reefs: structure, form and process. Springer, Dordrecht, pp 679–683

Frohlich C, Hornbach MJ, Taylor FW, Shen CC, Moala A, Morton AE, Kruger JAF (2009) Huge erratic boulders in Tonga deposited by a prehistoric tsunami. Geology 37:131–134. doi:10.1130/G25277A.1

Goff JR, Terry JP (2012) Living with natural hazards in the Asia–Pacific region. In: Terry JP, Goff JR (eds) Natural hazards in the Asia–Pacific region: recent advances and emerging concepts. Geological Society of London, Special Publication no. 361, London, pp 1–2. doi:10.1144/SP361.1

Martínez ML, Intralawan A, Vázquez G, Pérez-Maqueo O, Sutton P, Landgrave R (2007) The coasts of our world: Ecological, economic and social importance. Ecol Econ 63:254–272. doi:10.1016/j.ecolecon.2006.10.022

McKenna J, Jackson DWT, Cooper JAG (2011) In situ exhumation from bedrock of large rounded boulders at the Giant's Causeway, Northern Ireland: an alternative genesis for large shore boulders (mega-clasts). Mar Geol 283(1–4):25–35. doi:10.1016/j.margeo.2010.09.005

Morton RA, Richmond BM, Jaffe BE, Gelfenbaum G (2006) Reconnaissance investigation of Caribbean extreme wave deposits—preliminary investigations, interpretations, and research directions. Open-File Report 2006-1293. USGS. 41

Paris R, Fournier J, Poizot E, Etienne S, Morin J, Lavigne F, Wassmer P (2010) Boulder and fine sediment transport and deposition by the 2004 tsunami in Lhok Nga (western Banda Aceh, Sumatra, Indonesia): a coupled offshore–onshore model. Mar Geol 268:43–54. doi:10.1016/j.margeo.2009.10.011

Paris R, Naylor LA, Stephenson WJ (2011) Boulders as a signature of storms on rock coasts. Mar Geol 283:1–11. doi:10.1016/j.margeo.2011.03.016

Paris R, Wassmer P, Sartohadi J, Lavigne F, Barthomeuf B, Desgages E, Grancher D, Baumert P, Vautier F, Brunstein D, Gomez C (2009) Tsunamis as geomorphic crises: lessons from the December 26, 2004 tsunami in Lhok Nga, West Banda Aceh (Sumatra, Indonesia). Geomorphology 104:59–72. doi:10.1016/j.geomorph.2008.05.040

Richmond BM, Morton RA (2007) Coral-gravel storm ridges: examples from the tropical Pacific and Caribbean. In: Kraus NC, Rosati JD (eds) Coastal Sediments '07: Proceedings of the sixth international symposium on coastal engineering and science of coastal sediment processes, New Orleans, Louisiana, 13–17 May 2007, pp 572–583

Scheffers A, Scheffers S, Kelletat D, Browne T (2009) Wave-emplaced coarse debris and mega-clasts in Ireland and Scotland: boulder transport in a high-energy littoral environment. J Geol 117(5):553–573. doi:10.1086/656356

Shanmugam G (2012) Process-sedimentological challenges in distinguishing paleo-tsunami deposits. Nat Hazards 63:5–30. doi:10.1007/s11069-011-9766-z

Terry JP, Etienne S (2011) "Stones from the dangerous winds": reef platform mega-clasts in the tropical Pacific Islands. Nat Hazards 56(3):567–569. doi:10.1007/s11069-010-9697-0

Wassmer P, Baumert P, Lavigne F, Paris R, Sartohadi J (2007) Les transferts sédimentaires associés au tsunami du 26 décembre 2004 sur le littoral Est de Banda Aceh à Sumatra (Indonésie). Géomorphologie: Relief, Processus, Environnement 4:335–346. doi:10.4000/geomorphologie.4702

Chapter 2
Historical Review and Changing Terminology

Abstract Coastal boulders are often prominent features in the coastal landscape, sometimes mapped on nautical charts and named by local people. The vernacular names occasionally furnish vital clues to identify the original event that created the boulders. Coral boulders were first mentioned in the literature two centuries ago when scientists started to explore the Great Barrier Reef. The significance of coral boulders has long fuelled scientific debate, interpreted by some as remnants of former elevated reefs or by others as an inheritance from the action of past storm waves. For more than a century 'negro-head' was the seminal expression used to portray the emerged black-coloured rocks observed on reef platforms. Fortunately, this inappropriate choice became outmoded during the twentieth century and new terms have since been employed. Despite the existence of a specified grain-size scale to define large clasts, inconsistent nomenclature and the plethora of synonyms now in use causes some confusion.

2.1 Named Coastal Landmarks

Proof that people have recognised coastal boulders as an important feature of their local landscape is not difficult to find. In coastal locations where either individual boulders or clusters of boulders are prominent landmarks, or are well known for other reasons, it has been quite common for them to be given names by the local people. Of particular relevance in the present context are those cases where connections can be identified between boulders and high-energy marine inundation (HEMI) events, either known or unknown from the past. Examples have been documented in places as far-flung as Japan, Fiji and French Polynesia (Goto et al. 2010; Terry and Etienne 2010). Importantly, evident from several examples is that boulder names may reveal clues about the type of HEMI event that emplaced them. On Tupai Atoll in French Polynesia, for instance, a large coral boulder approaching 18 m^3 in size named *Paeotini* means 'thrown on the reef by the wind several times' in the local Polynesian

Fig. 2.1 *Vatu ni Cagi Laba* in Fijian, meaning 'stone from the strong wind'. These coral boulders sit on the Cakau Levu (Great Sea Reef), a barrier reef separated by an 8 km-wide lagoon from the northern coast of Vanua Levu Island in Fiji. According to local fishermen from Visoqo village, these boulders, amongst others, appeared on the reef after the passage of Tropical Cyclone Ami in January 2003. On the otherwise featureless reef flat, the boulders provide an important navigational aid to reach favourite fishing grounds. Photos by J. Terry, May 2009

dialect, which testifies to the boulder's storm origin. Similarly, on the Great Sea Reef 8 km off the north coast of Vanua Levu Island in Fiji, a conspicuous rock had been christened *Vatu ni Cagi Laba* by local fishermen (Fig. 2.1). Its name 'Stone from the strong wind' makes reference to the ferocity of Tropical Cyclone Ami in January 2003, which tossed it and other coral boulders nearby onto the reef. On Ishigaki Island in the Ryukyu Islands of southern Japan there are several so-called *tsunami-ishi*. Translating into English as 'tsunami stone', the *tsunami-ishi* are known to have been shifted from prior positions on the coast by the 1771 Great Meiwa tsunami. A selection of additional named boulders linked to HEMI events and information associated with them is given in Table 2.1.

Elsewhere, conspicuous coral boulders may be associated with local myths or legends, undocumented but remembered through storytelling as a way of passing down oral histories. Indeed, within the Asia–Pacific region in particular, traditional environmental knowledge (TEK), unwritten but disseminated through vernacular languages is an often overlooked ethno-geographical source of information on (prehistoric) geological disturbances and geohazards. According to local folklore, on Makin Island in the atoll nation of Kiribati in the central Pacific, two giant rocks called *Reuba* and *Tokia* (3°20.12′N 172°58.77′E, Fig. 2.2) are the result of two huge waves sent in anger by an ancient king residing on Butaritari Island, a few kilometres to the south, as a reprisal against the people of Makin for having supplied him with rotten breadfruit. Apparently, the two huge waves split the original Makin Island into separate smaller islets. Since the latitude of Makin is close enough to the Equator to be outside the tropical cyclone belt, and in the absence of other geomorphic processes in an atoll environment that might be responsible for

Table 2.1 Local names given to selected coastal boulders that were either emplaced or displaced by HEMI events

Location	Name of boulder(s) and meaning if available	Description or other points of interest[a]	Origin or cause of movement	Reference/data source
Vanua Levu Island, Fiji	*Vatu ni Cagi Laba* 'stone from the strong wind'	Size: 2.5 × 1.7 × 1.24 m Landmark for local fishermen. Situated on outer barrier reef, 8 km off the north coast from Visoqo village	Tropical Cyclone Ami, January 2003	Terry and Etienne (2010)
Tupai Atoll, French Polynesia	*Paeotini* 'thrown on the reef by the wind several times'	Size: 18 m³ Recently removed from the reef flat as perceived danger to airstrip	Unknown storms	Terry and Etienne (2010)
Yasura village, Ishigaki Island, Japan	*Yasura-ufukane* or *Ifangani* Latter is a local nickname meaning 'iron at Iha area'	Size: 6 × 3 × 3 m History: moved 54.6 m northwards by the Meiwa tsunami. Now located 1500 m from the reef edge	1771 Meiwa tsunami	Kawana (2003) cited by Goto et al. (2010)
Inoda village, Ishigaki Island, Japan	*Amatariya-Suuari* Suuari means 'rough sea' in the Ishigaki dialect	Size (2 separate boulders): 11 × 9 × 6 m; 6 × 5 × 5 m History: transported from the sea to the shore prior to 1771, but then moved landward by the Meiwa tsunami. Now located 1400 m from the reef edge at 10 m elevation	1771 Meiwa tsunami	Kawana (2003) cited by Goto et al. (2010)
Ohama village, Ishigaki Island, Japan	*Taka-Koruse ishi*	Size (2 separate boulders according to historical description): both approximately 7.3 m in length History: located in Koruse-Utaki Shrine before being displaced by the Meiwa tsunami One boulder (already split into several pieces) was identified >600 m from the reef edge at 5 m elevation), the other is not found	1771 Meiwa tsunami	Kawana (2003) cited by Goto et al. (2010)
Ishigaki Island, Japan	*Fukuraori-ishi*	Size (unconfirmed): 10.9 × 3.6 × 6.1 m History: originally located on Itokazu beach at Hirae village. Tsunami backwash displaced it to the reef flat in 1771 The boulder has not been identified in the area recently, possibly suggesting a renewed phase of movement	1771 Meiwa tsunami	Kawana (2003) cited by Goto et al. (2010)

[a] In the examples from the Ryukyu Islands of Japan, the written record of these named boulders can be traced back to at least the eighteenth century when boulder movements by the 1771 Meiwa tsunami were clearly described (Kawana and Nakata 1994, in Japanese, as cited in Goto et al. 2010)

Fig. 2.2 *Rebua* rock (shown above) and nearby *Tokia* rock are indicated on topographic maps of remote Makin Atoll in the Gilbert Islands group of Kiribati in the central Pacific, and stand out as especially conspicuous landmarks on the coastline of the low-lying atolls where the highest land barely reaches 2 m above sea level. Traditional stories explaining how the rocks were cast up by giant waves in ancient times make for interesting listening. A tsunami source for such waves seems likely since the atoll lies close to the Equator (3°N) and therefore outside the belt affected by tropical cyclones. See text for details. Photo courtesy of Emili Artack of SOPAC (Applied Geoscience and Technology Division of the Secretariat of the Pacific Community)

their emplacement, the presence of these giant boulders therefore suggest a tsunamigenic origin for the giant waves, possibly locally generated by a submarine landslide.

2.2 Earliest Scientific Observations of Coral Boulders

Complementary to indigenous environmental knowledge ingrained in local memory as mentioned earlier, it is noteworthy that coastal boulders have likewise been the focus of formal scientific attention by observers from various disciplinary fields over a considerable period of time. Accordingly, coastal boulders have been investigated for different reasons and from disparate points of view throughout history. In order to appreciate the interesting progression of coastal boulder studies, with a particular focus on reef-platform coral boulders (RPCBs), it is helpful to trace some of the influential accounts through time. The following review therefore highlights the changing emphasis that is apparent since the early nineteenth century.

Running parallel with shifting motivations for the scientific investigation of boulders has been a gradual evolution in the terminology used to refer to them.

With a documentation history spanning several centuries, 'coastal boulder' was neither the first nor the only term given to large rocks observed on coastlines. An array of different terms is found in older documents, but even within modern studies an inconsistent terminology is encountered. This situation poses obstacles when sifting through archival and published work to find mention of such deposits, because some accounts may possibly be overlooked owing to the variety of expressions used. For the purpose of clarification, changing terminology through time and that applied nowadays in coastal research will be discussed in conjunction with the historical summary below.

One of the earliest written descriptions of coral boulders was penned by the English navigator Captain Matthew Flinders. His book *A Voyage to Terra Australis* (Flinders 1814) was a journal recording a British sea expedition to the Australian continent in the years 1801–1803, when this great southern landmass was still largely unknown to Europeans. When arriving at the Great Barrier Reef (GBR) in October 1802, on several occasions Flinders noted the eye-catching coral features in the sea, which he referred to as 'negro heads'[1] (see Box 1). As described by Flinders (1814), these rocks were dry and black, covered during high water, within which corals and shells could clearly be distinguished. In one entry he mentioned that no further description of form and position was needed as this information could be learnt from the chart. This implies that these rocky protrusions on the surface of reefs were deliberately marked on the navigational charts that were being drawn up during this voyage of exploration, most likely because of the danger they posed to sailing. From his writings, it is clear that Flinders understood the fragments of coral rock, originally white in colour, had become blackened through exposure to the elements on reef surfaces. British sailors in the late eighteenth century commonly used 'negro head' to refer to exposed coral masses that were black in appearance, so this is probably the reason why Flinders followed suit and adopted the same expression in his journal.

---

[1] To all modern readers this antiquated expression should rightfully be considered as repugnant and wholly inappropriate. As such, its usage is in no way condoned by the authors nor is any offense intended. The expression is mentioned solely for the purpose of providing an accurate historical review, since this was the recognised term for reef-platform boulders for well over a century.

Box 1
Matthew Flinders' Observations of Coral Boulders on the
Great Barrier Reef in 1802

A *Voyage to Terra Australis* (1814) by Matthew Flinders was one of the earliest publications that mentions coral boulders on reefs. During his circumnavigation of Australia in 1802 as commander of HM Sloop *Investigator*, the distinguished navigator and cartographer used the term 'negro heads' for the conspicuous rocks he saw in various locations on the Great Barrier Reef: Tuesday 5th Oct 1802:

> The reefs were not dry in any part, with the exception of some small black lumps, which at a distance resembled the round heads of negroes; the sea broke upon the edges, but within side the water was smooth, and of a light green colour. A further description of these dangers is unnecessary, since their forms and relative positions, so far as they could be ascertained, will be best learned from the chart.

Friday 8th October 1802:

> We seemed at this time to be surrounded with reefs; but it was ascertained by the whale boat, that many of these appearances were caused by the shadows of clouds and the ripplings and eddies of tide, and that the true coral banks were those only which had either green water or negro heads upon them. Of these, however, there was a formidable mass, all round a-head, with but one small channel through them;

Saturday 9th Oct 1802:

> Different corals in a dead state, concreted into a solid mass of a dull-white colour, composed the stone of the reef. The negro heads were lumps which stood higher than the rest; and being generally dry, were blackened by the weather; but even in these, the forms of the different corals, and some shells were distinguishable.

[excerpts from Flinders (1814), vol. 2, pp. 83–88].

2.3 The Eruption of Krakatau Volcano in 1883

In August 1883, the cataclysmic volcanic eruption of Krakatau Island in Indonesia generated global shock waves and a tsunami that was felt as far away as New Zealand and South Africa. In the aftermath of the devastating inundation experienced on the west coast of Java, which washed away entire towns, eyewitnesses described how numerous blocks of coral encumbered roads and were strewn far inland. One report was communicated by Neale (1885) after he visited Merak in northwest Java six weeks after the tsunami had struck. Neale recorded that numerous coral fragments of immense size were found deposited on land two–three miles (3–5 km) from the sea. The detailed nature of his narrative indicates that the ability of giant tsunami waves to transport coral fragments of considerable size was clearly recognised:

> But one of the most remarkable facts concerning the inundation remains to be told. As we walked or scrambled along we were much surprised to find great masses of white coral rock

lying at the side of our path in every direction. Some of these were of immense size, and had been cast up more than 2 or 3 miles from the seashore. It was evident, as they were of coral formation, that these immense blocks of solid rock had been torn up from their ocean bed in the midst of the Soenda Straits [Sunda Straits], borne inland by the gigantic wave, and finally left on the land several miles from the shore. Anyone who had not seen the sight would scarcely credit the story. The feat seems an almost impossible one. How these great masses could have been carried so far into the interior is a mystery, and bears out what I have said in previous papers as to the height of this terrible wave. Many of these rocks were from twenty to thirty tons in weight, and some of the largest must have been nearly double. Lloyd's agent, who was with me, agreed in thinking that we could not be mistaken if we put down the largest block of coral rock that we passed, as weighing not less than fifty tons (Neale 1885, pp. 486–488, 545–557, 635–638; as recounted in Simkin and Fiske 1983, p. 121).

Of special interest was the discovery at Anyer (6°03′S 105°55′E) of a block of gargantuan proportions: 6.5 m high, with a volume of 300 m^3 and estimated weight approaching 600 tons (Fig. 2.3). Astonishingly, this enormous clast was found 100 m inland from the coast and aroused such levels of curiosity that it was carefully measured, mapped, photographed and reported in early Dutch accounts of the Krakatau eruption (Verbeek 1885).

2.4 Reef Remnants Versus Storm Deposits: Competing Ideas, Early 1900s

Following the earliest notes on reef-platform coral boulders by Matthew Flinders during his southern voyage of exploration, from the 1890s onwards coral boulders on the Great Barrier Reef (GBR) once again became a focus of interest, but this time by scientists who were concerned with trying to understand the growth and long-term development of coral reefs. Alexander Agassiz was one of the pioneers who investigated the GBR for two months in 1896 and later published his ideas on reef formation (Agassiz 1898, 1903). Agassiz proposed that bouldery deposits should be considered as the last remnants of an old reef that had grown up at an earlier period of higher sea level and had then been left behind as surface remnants by the processes of weathering and erosion (Agassiz 1898; Hopley et al. 2007). His opinion was shared by David and Sweet (1904, p. 71) who described similar features on Funafuti Atoll (formerly Ellice Islands, now Tuvalu): "We agree with Professor Agassiz's general opinion that such "negro-heads" are of the nature of outliers, and do not represent once loose blocks of reef limestone cast up upon the reef platform by storms and subsequently cemented down. Our observations at Funafuti led us to the conclusion that these "negro-heads" indicate downward movement of the shore line." However, these authors stated in a footnote on the same page how "Messrs. Halligan and Finckh consider that these blocks are not in situ, and they do not therefore look upon them as evidence of such a movement of the shore line". These notes testify to the emerging questions and contradictory views about the significance and interpretation of reef-top boulders.

The subsequent decades proved to be a time of critical thinking and new theories regarding Quaternary coral reef formation, and so the presence of coral

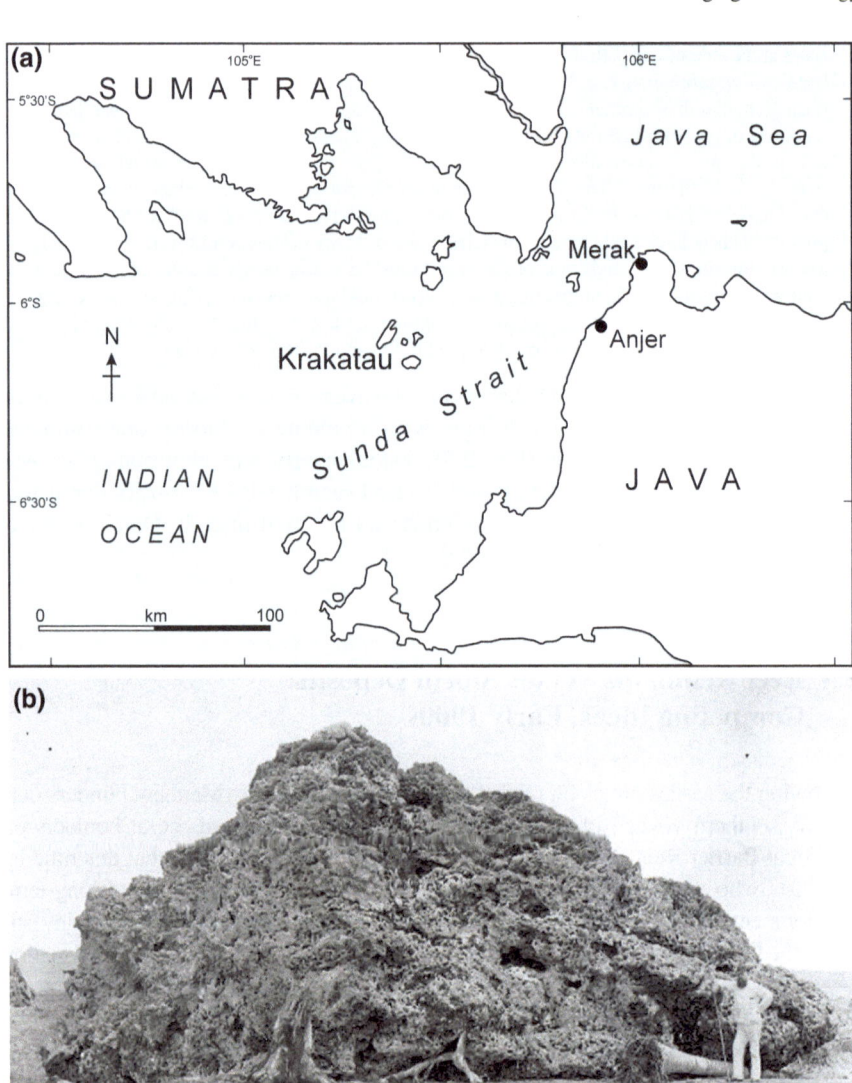

Fig. 2.3 a Krakatau Island in the Sunda Strait (Indonesia) where the infamous volcanic eruption of 1883 generated a global tsunami. **b** 1885 photograph of the 300 m^3 coral block deposited at Anyer (Anjer) on the west Javanese coast. Original photo credit: Woodbury and Page, *circa* 1885, held at the archives of the Royal Tropical Institute, Amsterdam, The Netherlands (http://collectie. tropenmuseum.nl/default.aspx?idx=ALL&field=*&search=60005541)

boulders was discussed within the scientific debates of the time. Boulders on coral reefs were consequently examined as an important component of coastal sedimentology and several prominent researchers began to associate their occurrence with past storm activity. Of these, most continued to use the terminology for reef boulders that Flinders had introduced a century earlier (see next section).

One of the most significant contributions was that of Hedley and Taylor (1907) who conducted a major study of the structure of the GBR. Their report was particularly insightful as it described not only the form, size, distribution, orientation and composition of coral boulders, but also suggested that their longevity might be relatively limited. This suggestion was based on estimated rates of clast erosion (2 inches [5 cm] over 4–5 years) determined from the protrusion of old oyster encrustations. The authors also firmly rejected Agassiz' notion that coral boulders represented denuded fragments of ancient elevated reefs. Hedley and Taylor wrote such a richly informative narrative of the coral boulders they observed on Cairns Reef that a summary would do an injustice; therefore the greater part of the original account is given below:

> Sailing past almost any reef on the Barrier the traveller's attention is arrested by a row of crags, like huge milestones, irregularly disposed along the crest of the reef. Even where the bank is covered with high water they project above the surface of the sea. To these Flinders applied the name "negroheads,"… If the negroheads gathered on a steep and narrow shore they would compose a "hurricane beach."

> The following description is based on those observed on Cairns Reef. At a distance of a mile or so, by optical illusion, perhaps partly mirage, partly lack of standards of comparison, the negroheads are accepted by the eye as being of far larger bulk than the reality. On near approach they resolve into masses of dead coral 5ft. or 6ft. in height and of nearly equal breadth.

> The massive corals, *Porites*, *Astrea*, &c., of which they are composed, grow irregularly and circumferentially. It is impossible to orient them by any axis of growth or to tell by inspection of the polyp cells whether they are upright or upset. Once landed, a stranded block might be welded by chemical action to the floor on which it stood. No sign of coral breccias or superposed coral was noticed in connection with the Cairns negroheads.

> No blocks occur on the central or leeward portion. The whole crop are confined to a zone 300yds. or 400yds. from the surf. Often the grit-armed surf has hollowed a pool around the boulder and undercut the base, leaving a stout stalk by which it is attached to the reef. Above the block may be fantastically carved into pinnacles and hollows. All are obviously melting away under rapid erosion and possibly solution.

> A time measure is afforded by encrusting oysters. These cluster thickly on the exterior of negroheads, and may even completely sheath in armor a square foot of the surface. Single aged oysters project like spurs, their point of attachment being *not the umbo* but the ventral margin.

> Their history clearly is that the oyster was first fastened as usual by the umbo, and that as soon as a fresh grip was gained in front so the earlier support behind was removed. During the brief span of the bivalve's life (say four or five years) a layer as much as 2in. thick of the rock crust may have vanished. At this rate of erosion no great antiquity can be ascribed to the negroheads.

> Both Kent and Agassiz have illustrated and described the Queensland negroheads, but offer opposing explanations of their origin. The former regards them as jetsam flung up by hurricanes; the latter considers them as a residue of elevated reefs cut down by erosion to present level. The verdict is a matter of some geological importance, for if the view of Agassiz be adopted a direct proof of recent alteration of level is established.

The hypothesis of Kent is preferred by us on the following grounds :—Positively—The negroheads do not continue down into the ground but are perched as morainic blocks might be. Jetsam would accumulate on the weather side of the reefs (where the negroheads are) not on the lee side (where they are absent). Negatively—An elevated reef in course of denudation would commence to wear on the windward side, where the attack is fiercest; the last surviving remnants should be on the leeward shore. Supposing that the negroheads are such remnants, why do they survive only where they ought earliest to disappear ? The central portions, more than half a mile from either edge, might naturally be expected to remain as more or less solid "mesas" long after the rest has been ground to sand. Such is not the case on Cairns Reef...." (Hedley and Taylor 1907, pp. 402–4).

Two decades later, Lenox-Conyngham and Potts (1925, p. 317, Fig. 2.4) summed up the divergent opinions about the origin of coral boulders held at the time, but themselves supported the idea that they were "thrown up on top of the reef during a storm and carved into their present shape by tidal erosion". Part of their reasoning was that boulders were always seen "on the weather side and not the lee side of a reef". Nathan (1927, p. 548) also noted that, in the northern region of the GBR, coral boulders are not present on the side of reef islands where broken coral fragments normally accumulate, but occur on the sides of islands that are most affected by hurricanes. He concluded that "the theory that these nigger heads are the last remains of a reef elevated above the present sea-level is, I believe, no longer seriously held". Similarly, Close et al. (1929, p. 268) proposed that "cyclones come and cause the big seas that break off masses of the coral and throw them on to the land already formed", while Steers (1929) claimed "the fact that these masses occur on the windward edges of reefs and also on the northwestern sides, from which direction heavy weather often comes, leaves little doubt as to their origin. Without question they are the products of wave action and merely represent the heavier material cast up". Two decades later, Fairbridge and Teichert (1948) explained that RPBCs and coral ramparts seen on the Low Isles of the GBR were all part of cyclonic deposits. Thus, by the middle of the twentieth century it was generally accepted among the greater body of commentators that coral boulders found on reef platforms were the product of storm activity, in particular tropical cyclones or similar high-energy weather systems.

It is important not to overlook investigations during the same period of coastal boulders comprising rock types other than coral limestone, which were also being recognised as storm deposits in locations outside tropical waters. A particular case stands out at Ben Buckler's Point near Bondi Bay, Sydney, in south eastern Australia. In July 1912 a violent storm was recorded to have thrown up a joint-bounded sandstone block measuring 26 × 20 × 10 feet (7.9 × 6.1 × 3 m) and weighing approximately 235 tons. The block was carried up from sea level onto an elevated platform at least 10 feet (3 m) high and transported horizontally for 160 feet (48.8 m) by the storm waves (Sussmilch 1912, Fig. 3.2). Sussmilch's report is especially notable as it marks the first attempt at calculating the lifting power required to move a block as a way of estimating, post-event, the energy of the storm waves. This quantitative approach to the analysis of storm-wave energy by using the dimensions and position of displaced boulders as proxy measurements has become a major theme in present-day coastal research, as is discussed in Chap. 3.

Fig. 2.4 From their observations on the Great Barrier Reef of Australia, Lenox-Conyngham and Potts (1925, pp. 324–325) concluded that large stranded masses of coral "testify to the force of the waves" and are therefore storm-derived deposits. Photo from Lenox-Conyngham and Potts (1925, p. 325)

2.5 Varying Expressions for Coral Boulders

Expressions like 'negro heads' and variants thereof continued to be widely employed by scientists for reef-platform coral boulders throughout the early decades of the twentieth century (e.g. Hedley and Taylor 1907; Lenox-Conyngham and Potts 1925; Nathan 1927; Steers 1929; Close et al. 1929, 1930). There were those, however, who were quite opposed to such terminology, although it must be said this was not in response to any revulsion felt by using such offensive terms, but rather because of the confusion brought about by their dual (geomorphic) meanings. As part of a study of island-reef systems on the Queensland coast of Australia, Spender (1930) returned to the original writings of Flinders and noted how terminological confusion was caused by 'negro heads' having been used to refer to two distinct types of reef features: coral pinnacles with their tops exposed above sea level that are still attached at their base to the reef framework (Fig. 2.5) and fragments of dead coral material lying on the reef surface that may or may not become cemented in place. Consequently, Spender suggested abandoning the standard terms and favoured adopting 'coral-head' for coral pinnacles and 'boulders' for detached coral masses resting on reef platforms:

> In passing it is worth making a comment on the use of the word "nigger-head" or "negro-head"; at present both in the literature and general usage it seems to have two meanings. Usually it has been applied to the large boulders which stand, often isolated, on the surface of reefs; but it has also been used to describe the coral heads of the above paragraph.
>
> Reading Flinders' definition recalls to me the occasional skull-shaped colony of such a coral as Favia or Symphyllia, recently dead and blackened, standing out from the top of a coral head… However, by the terminological confusion which has taken place, scientists have called them nigger-heads while they lived in the deep water and nigger-heads again while they lay decaying on the reef surface, riddled by a thousand boring organisms. The circumstances seem to justify abandoning the word for the present, and using the words "coral-head" and "boulder" for the two meanings respectively. (Spender 1930, p. 203).

In spite of this recommendation by Spender, later Fairbridge and Teichert (1948) indicated their preference to stick with the long-standing 'negro-head' for

Fig. 2.5 Tikehau Atoll in French Polynesia: the base of jagged coral pinnacles remain attached to the modern reef structure in their original position of growth and should not be mistaken for detached reef boulders. Coral pinnacles are vestigial remnants of an emerged reef surface that was formed at some time in the past when sea level was above its current position. Across the Tuamotu Archipelago such pinnacles are common. Known as *feo* in local Polynesian languages (Nunn 1994), they are testimony to a mid-Holocene sea-level highstand in the central Pacific. Where the pinnacles have a narrow base due to marine erosion and resemble the shape of mushrooms, they are known as *rochers-champignons* [a French name, meaning 'mushroom rocks' (Ottmann 1962)]. Photo by S. Etienne, June 2011

detached coral boulders. What is apparent is that 'coral head' is also a problematic term, since coral heads may be found in different settings, either submerged and living, or exposed and dead:

> We prefer to adhere to the widely-used term "negro-head" as indicating the largest-sized boulder of dead coral which is commonly found on top of exposed reef crests. Pinnacles of living coral below low tide may be known as "coral-heads". In Queensland, both of these completely different features are referred to colloquially as "niggerheads".

> There seems to be nothing in this [Flinders'] definition that might lead to confusion with submerged living "coral heads", nor is there any reason to follow the example of those who have changed the term to "niggerheads" (Fairbridge and Teichert, 1948, p. 81).

Eventually, it was Newell (1954) who made an unambiguous statement on the offensive nature of the early terms, and instead coined the new phrase 'reef block':

> In spite of a natural repugnance of the inelegant term 'nigger-head' or 'negrohead' almost universally employed by students for these reef blocks, I would follow accustomed usage if there were justification on grounds of special aptness. There is none. Let us call a reef block a reef block" (Newell 1954 p.32, Fig. 2.6) .

Thankfully, since the middle of the twentieth century, 'negro head' and similar offensive expressions became increasingly unpopular and eventually fell into disuse in the mainstream geomorphology literature, apart from rare exceptions (e.g. Guilcher 1988). Outside the English-speaking scientific community, however, French geomorphic vocabulary saw the introduction of *têtes de nègre*, the direct translation of 'negro heads', by André Guilcher in 1950 (Figs. 2.7 and 2.8). Francophone researchers have widely adopted this phrase for coral boulders on reefs and reef islands, the use of which has lingered on until relatively recent times

Fig. 2.6 Large reef block observed on Mataira Islet, Raroia Atoll in the Tuamotu Archipelago of French Polynesia. The block at the right is approximately 10 m long and the volume estimated as c.255 m³. *Source*: Newell (1954), with permission

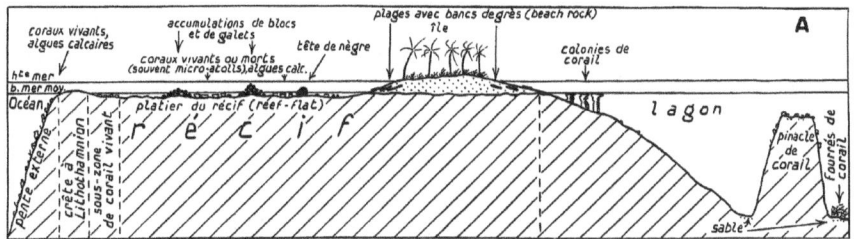

Fig. 2.7 The introduction of the term *têtes de nègre* in French geomorphic vocabulary by Guilcher (1950, p. 184), a diagram inspired by Dryden (1944)

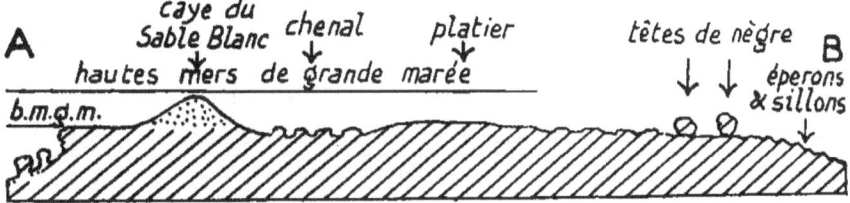

Fig. 2.8 *Têtes de nègre* on a reef near Saziley du Sud, Mayotte in the Indian Ocean. *Source*: Guilcher et al. (1965, p. 73). The authors estimated the size of the coral boulders as 2 × 1.3 m

(e.g. Guilcher et al. 1965; Blanc et al. 1966; Battistini 1970, 1978; Bourrouilh-Le Jan and Talandier 1985; Paskoff 1993; Bourrouilh-Le Jan 1994; Guillaume et al. 1997).

2.6 Perspectives on Sediment Clast Size

In the field of sedimentology, a standard system to describe the size of sedimentary particles has been in existence since the development of the Udden-Wentworth (U-W) grain-size scale almost a century ago (Udden 1914; Wentworth, 1922, 1935). However, in the original scale, only grains smaller than 4096 mm were classified. It was not until 1999 that Blair and McPherson extended the scheme to include coarser sediments with clasts of intermediate axis (i.e. b-axis) length up to 1075 km. Extension of the existing scale was preferred over other possible alternatives, such as using lively expressions of comparison like "boulders the size of

Volkswagens" (Blair and McPherson 1999, p. 6). On the extended U-W scale, clasts between 0.25 and 4.1 m are classified as 'boulders', clasts 4.1–65.5 m are 'blocks'. (The length of the intermediate axis is used as its measurement is equivalent to sieve analysis of fine-grained particles). Still larger clasts are classified as 'slabs', 'monoliths' or 'megaliths', according to the ranges shown in Table 2.2. All classes bigger than the boulder category are collectively referred to as 'megaclasts'.

Although the extended U-W particle-size scale is now adequate to cover very sizeable clasts, several terminological difficulties are still apparent in the coastal literature. First of all, the term 'boulder' suggests a certain geometry denoting an element of roundness, whereas the term 'block' implies a degree of angularity or 'blockiness' of form, regardless of true shape (or size). A more pervasive problem, however, is that the clearly-defined term boulder is frequently applied incorrectly, beyond its properly assigned size range on the U-W scale. Oak (1984) for instance cited published descriptions of so-called 'boulder beaches' that in reality comprised cobble-grade material. Of course, it is easy to appreciate why this is the case, because in addition to its sedimentological application *sensu stricto*, the word 'boulder' also enjoys wider usage in colloquial language. This situation arises because an easily recognisable name is needed in a general sense to mean a large weather-worn or water-worn stone, and 'boulder' seems to meet the requirements better than any alternative that instantly comes to mind. Indeed, for convenience the expression 'coastal boulder' is used in this book in a generic way to represent all such deposits, and also because the majority of clasts described in the published literature do in fact fall into the boulder-size category in the revised U-W system.

Table 2.2 The grain-size scale for sedimentary particles of Blair and McPherson (1999)

Class	Grade	Particle length (intermediate axis)	
		ϕ units	Metric units
Megalith	(5 grades)	−25 to −30	33.6–1075 km
Monolith	(5 grades)	−20 to −25	1–33.6 km
Slab	(4 grades)	−16 to −20	65.5–1048.6 m
Block	Very coarse	−15 to −16	32.8–65.5 m
	Coarse	−14 to −15	16.4–32.8 m
	Medium	−13 to −14	8.2–16.4 m
	Fine	−12 to −13	4.1–8.2 m
Boulder	Very coarse	−11 to −12	2.0–4.1 m
	Coarse	−10 to −11	1.0–2.0 m
	Medium	−9 to −10	0.5–1.0 m
	Fine	−8 to −9	0.25–0.5 m
Cobble	(2 grades)	−6 to −8	64–256 mm
Pebble	(4 grades)	−2 to −6	4–64 mm
Granule	(1 grade)	−1 to −2	2–4 mm
Sand	(5 grades)	4 to −1	0.063–2 mm
Silt	(4 grades)	8 to 4	0.004–0.063 mm
Clay		<8	<0.004 mm

Particle length is based on the intermediate axis (b-axis). The table has been expanded to show the size ranges of the four grades within the 'boulder' and 'block' categories

Table 2.3 A selection of the terms used for coastal boulders in research publications

Term used or mentioned	Example of publication	Study area	Size of clast concerned/note
Block	Simkin and Fiske (1983)	Anyer, Java coast	300 m^3, height 6.5 m, 600 tons
	Yu et al. (2004)	Nansha area, South China Sea	1–2.5 m in height
	Leroy (2012)	Between Anyer and Carita, Java	No dimensions given but included a photo of the block
Boulder and Megaclast	Lorang (2011)	Kalalau Beach, Hawaii	Developed mathematical equations from previous studies, with a conceptual example from the study area
Coarse debris and Megaclasts	Scheffers et al. (2009)	Ireland and Scotland	Up to >250 tons
Bloc cyclopéen	Bourrouilh-Le Jan and Talandier (1985)	Rangiroa, French Polynesia	Largest size: 15 × 10 × 5 m
	Fichaut and Suanez (2008)	Banneg Island, Brittany, France	Largest size: 5.3 × 3.9 × 0.5 m
Mega-boulder	Kelletat et al. (2007)	West coast of Thailand	"Mega-boulder is no set term" (p. 423) Boulders from 50 tons (indicate size by weight)
Megablock	Frohlich et al. (2011)	—(Encyclopedia entry for megablock)	Lists 'boulders or megaboulder deposits' as synonyms of megablocks (p. 679)
Megaclast	Noormets et al. (2002, 2004)	Oahu, Hawaii	Largest weight 96 tons (8.8 × 5.5 × 2.5 m)
	Williams and Hall (2004)	North Atlantic	Largest size 6 × 5 × 3.2 m

Apart from the nomenclature already mentioned, the additional term 'megablock' has appeared as a *de facto* synonym for very large clasts, especially those found on coral reefs. Frohlich et al. (2011, p. 679) define megablocks as "intact blocks or boulders, often composed of coral and occasionally with dimensions of 10 m or greater". Elsewhere, the colourful phrases 'cyclopean blocks', and the French equivalent *blocs cyclopéens*, have also been applied to large coastal clasts, but without any precise designation of size, which may create confusion. *Blocs cyclopéens* was first coined by Hallégouët 1982 in a article to describe coarse deposits on Banneg Island off the Brittany coast in north west France (Fichaut and Suanez 2008) and was subsequently used in other scientific journals from 1984 onwards (e.g. Hallégouët 1984; Bourrouilh-Le Jan and Talandier 1985). The expression makes reference to the cyclopean stones (i.e. stones so enormous that only the mythical giant Cyclops had the strength to lift them) that the ancient Greeks used as masonry to build fortifications at Mycenae (Fichaut and Suanez 2008). *Blocs cyclopéens* was then translated into the English equivalent 'cyclopean blocks' by Ricard (1985) and was again used by Hearty (1997) when referring to earlier work of French authors in Polynesia. Since then, both French and English versions appear in various coastal literature until today, particularly that which focuses on HEMI events (e.g. Etienne 2007; Fichaut and Suanez 2008; Ardhuin et al. 2011).

Partly in response to the issues outlined above, Paris et al. (2011) carried out a critical review of coastal boulder studies across different environmental regimes. They observed that most of the coarse-grained clasts investigated in coastal sedimentological studies are actually boulder-sized, while some fall within the fine-block size range (4.1–8.2 m). Moreover, no clast with intermediate axis (or diameter) greater than 8.2 m (the upper limit for fine block) was reported among the publications they reviewed. Table 2.3 provides an overview of the various terms used specifically for HEMI-emplaced coastal boulders, with some examples of publications. It is clear that current terminology remains inconsistent. Consequently, Paris et al. (2011) recommended that researchers of coastal deposits should be more vigilant in future and follow the modified U-W grain-size scheme of Blair and McPherson (1999) in a disciplined manner. Furthermore, researchers should explicitly avoid misusing the word 'megaclast', as this denotes a specific range of very large clast sizes that are normally too big to be present in coastal settings.

References

Agassiz A (1898) A visit to the Great Barrier Reef of Australia in the steamer Croydon during April and May, 1896. Bull Mus Comp Zool Harv Coll 28:95–148

Agassiz A (1903) The coral reefs of the tropical Pacific. University Press, Cambridge

Ardhuin F, Pineau-Guillou L, Fichaut B, Suanez S, Corman D, Filipot JF (2011) Extreme set-up and run-up on steep cliffs (Banneg Island, France). In: Actes de colloque 12th international workshop on wave hindcasting and forecasting, and 3rd coastal hazard symposium, Kohala Coast, Hawai'i, HI, 2011, p 9

Battistini R (1970) Etat des connaissances sur la géomorphologie de l'île Maurice. Revue de Géographie de Madagascar 17:63–77

Battistini R (1978) Les récifs coralliens de la Martinique. Comparaison avec ceux du sud ouest de l'Océan Indien. Cahiers de l'ORSTOM série Océanographie 16(2):157–177

Blair TC, McPherson JG (1999) Grain-size and textural classification of coarse sedimentary particles. J Sedim Res 69:6–19. doi:10.1306/D426894B-2B26-11D7-8648000102C1865D

Blanc JJ, Chamley H, Froget C (1966) Sédimentation paralique et récifale à Tuléar. Annales de l'Université de Madagascar, série Sciences Naturelles et Mathématiques 4:35–79

Bourrouilh-Le Jan FG (1994) Les récifs coralliens: indicateurs de l'environnement et des paléoenvironnements, In: Maire R, Pomel S, Salomon JN (eds), Enregistreurs et indicateurs de l'évolution de l'environnement en zone tropicale. Press Univ Bordeaux, Talence, pp 275–297

Bourrouilh-Le Jan FG, Talandier J (1985) Sédimentation et fracturation de haute énergie en milieu récifal: tsunamis, ouragans et cyclones et leurs effets sur la sédimentologie de la géomorphologie d'un atoll: motu et hoa, à Rangiroa, Tuamotu, SE Pacifique. Marine Geol 67:263–333. doi:10.1016/0025-3227(85)90095-7

Close C, Nathan M, Gardiner JS, Bidder GB, Steers JA (1929) The Queensland coast and the Great Barrier Reefs: discussion. Geogr J 74(4):367–370

Close C, Nathan M, MacArtney EH, Steers JA, Dr Stephenson, Spender M (1930) Island-Reefs of the Queensland Coast : discussion. Geogr J 76(4):294–297

David TWE, Sweet G (1904) The Geology of Funafuti. In: Bonney TG (ed) The Atoll of Funafuti: borings into a coral reef and the results. Royal Society of London, London, pp 61–88

Dryden AL (1944) Surface features of coral reefs. U.S. beach erosion board. Technical Report, 4, p 62

Etienne S (2007) Les plates-formes d'érosion marine des littoraux volcaniques. In: Etienne S, Paris R (eds) Les littoraux volcaniques, une approche environnementale. PUBP, Clermont-Ferrand, pp 37–55

Fairbridge RW, Teichert C (1948) The Low Isles of the Great Barrier Reef: a new analysis. Geogr J 111(1):67–88

Fichaut B, Suanez S (2008) Les blocs cyclopéens de l'île de Banneg (archipel de Molène, Finistère): accumulations supratidales de forte énergie. Géomorph Relief, Process, Envir 1:15–32. doi:10.4000/geomorphologie.5793

Flinders M (1814) A voyage to Terra Australis, vol 2. Nicol, London, pp 83–88

Frohlich C, Hornbach MJ, Taylor FW (2011) Megablocks. In: Hopley D (ed) Encyclopedia of modern coral reefs: structure, form and process. Springer, Netherlands, Dordrecht, pp 679–683

Goto K, Kawana T, Imamura F (2010) Historical and geological evidence of boulders deposited by tsunamis, southern Ryukyu Islands, Japan. Earth-Sci Rev 102(1–2):77–99. doi:10.1016/j.earscirev.2010.06.005

Guilcher A (1950) Les récifs coralliens: formes et origines. L'Information Géographique 14(5):183–196

Guilcher A (1988) Coral reef geomorphology. Wiley and sons, Chichester

Guilcher A, Berthois L, Le Calvez Y, Battistini R, Crosnier A (1965) Les récifs coralliens et le lagon de l'ile Mayotte (Archipel des Comores, Océan Indien). ORSTOM, Paris

Guillaume M, Dauvin JC, Doumenc D (1997) Typologie des ZNIEFF-MER – Liste des milieux marins et des biocénoses marines des côtes françaises dans les D.O.M. MNHN, Paris

Hallégouët B (1982) Géomorphologie de l'archipel de Molène. Penn ar Bed 110:83–97

Hallégouët B (1984) Contribution à l'étude morphologique de l'archipel de Molène (Finistère). Études géographiques sur la Bretagne et questions diverses. In : Actes du 107ᵉ Congrès National des Sociétés Savantes, Brest 1982, Secteur de Géographie, CTHS, Paris, pp 61–77

Hearty PJ (1997) Boulder deposits from large waves during the last interglaciation on North Eleuthera Island, Bahamas. Quat Res 48(3):326–338. doi:10.1006/qres.1997.1926

Hedley C, Taylor TG (1907) Coral reefs of the Great Barrier, Queensland. A study of their structure, life-distribution, and relation to mainland physiography. Australia and New Zealand association for the advancement of science. Rep. Meeting 1907. Proceedings of Section C, pp 397–413

Hopley D, Smithers SG, Parnell K (2007) The geomorphology of the Great Barrier Reef: development, diversity and change. Cambridge University Press, Cambridge

Kawana T (2003) Chapter 5: Evaluation of the tsunami damage at Okinawa prefecture with a focus on the 1771 Meiwa tsunami, In: Research Institute for Subtropics (Ed.), Fundamental studies for the natural disaster risk and the countermeasure in Okinawa Prefecture, pp 263–328 (in Japanese, original title translated)

Kawana T, Nakata K (1994) Timing of late holocene tsunamis originated around the southern Ryukyu Islands, Japan, deduced from coralline tsunami deposits. Jpn J Geogr 103:352–365 (in Japanese with English abstract)

Kelletat D, Scheffers SR, Scheffers A (2007) Field signatures of the SE-Asian mega-tsunami along the west coast of Thailand compared to Holocene paleo-tsunami from the Atlantic region. Pure Appl Geophys 164(2–3):413–431. doi:10.1007/s00024-006-0171-6

Lenox-Conyngham G, Potts FA (1925) The Great Barrier Reef. Geogr J 65(4):314–329

Leroy SAG (2012) Natural hazards, landscapes, and civilizations. In: Shroder J Jr, James LA, Hardon C, Claque J (eds) Treatise on geomorphology, vol 13. Academic Press, San Diego

Lorang MS (2011) A wave-competence approach to distinguish between boulder and mega-clast deposits due to storm waves versus tsunamis. Mar Geol 283(1–4):90–97. doi:10.1016/j.margeo.2010.10.005

Nathan M (1927) The Great Barrier Reef of Australia. Geogr J 70(6):541–551

Neale P (1885) The Krakatoa Eruption. Leasure Hour 34:348–351, 379–388, 544–557, 635–638 (Printed also in Living Age 166:693, 753, 819; 167:174)

Newell ND (1954) Reef and sedimentary processes of Raroia. Atoll Res Bull 36:1–35

Noormets R, Felton EA, Crook KAW (2002) Sedimentology of rocky shorelines: 2. Shoreline megaclasts on the north shore of Oahu, Hawaii- origins and history. Sed Geol 150(1–2): 31–45. doi:10.1016/S0037-0738(01)00266-4

Noormets R, Crook KAW, Felton EA (2004) Sedimentology of rocky shorelines: 3. Hydrodynamics of megaclast emplacement and transport on a shore platform, Oahu, Hawaii. Sed Geol 172:41–65. doi:10.1016/j.sedgeo.2004.07.006

Nunn PD (1994) Oceanic Islands. Blackwells, Oxford

Oak LH (1984) The boulder beach: a fundamentally distinct sedimentary assemblage. Ann Assoc Am Geogr 74(1):71–82

Ottmann F (1962) L'atol das Rocas dans l'Atlantique sub-tropical. Revue de Géologie Dynamique et Géographie Physique 5:101–106

Paris R, Naylor LA, Stephenson WJ (2011) Boulders as a signature of storms on rock coasts. Mar Geol 283:1–11. doi:10.1016/j.margeo.2011.03.016

Paskoff R (1993) Les littoraux. Impacts des aménagements sur leur évolution. Masson, Paris

Ricard M (1985) Rangiroa atoll, Tuamotu archipelago. In: Delesalle B, Galzin R, Salvat B (eds) Fifth International Reef Congress, Tahiti, vol 1, pp 159–210

Scheffers A, Scheffers S, Kelletat D, Browne T (2009) Wave-emplaced coarse debris and mega-clasts in Ireland and Scotland: boulder transport in a high-energy littoral environment. J Geol 117(5):553–573. doi:10.1086/600865

Simkin T, Fiske RS (1983) Krakatau, 1883: the volcanic eruption and its effects. Smithsonian Institute Press, Washington, D.C.

Spender M (1930) Island-reefs of the Queensland coast. Geogr J 76(3):193–214

Steers JA (1929) The Queensland coast and the Great Barrier Reefs. Geogr J 74(3):232–257

Sussmilch CA (1912) Note on some recent marine erosion at Bondi. J Proc R Soc N.S.W. 46:155–158

Terry JP, Etienne S (2010) "Stones from the dangerous winds": reef platform mega-clasts in the tropical Pacific Islands. Nat Hazards 56(3):567–569. doi:10.1007/s11069-010-9697-0

Udden JA (1914) Mechanical composition of clastic sediments. Geol Soc Am Bull 25:655–744

Verbeek RDM (1885) Krakatau. Landsdrukkerij, Batavia, 495 pp

Wentworth CK (1922) A scale of grade and class terms for clastic sediments. J Geol 30:377–392

Wentworth CK (1935) The terminology of coarse sediments. Natl Res Counc, Bull 98:225–246

Williams D, Hall AM (2004) Cliff-top megaclast deposits of Ireland, a record of extreme waves in the North Atlantic–storms or tsunamis? Mar Geol 206(1–4):101–117. doi:10.1016/j.margeo.2004.02.002

Yu K, Zhao J, Collerson KD, Shi Q, Chen T, Wang P, Liu T (2004) Storm cycles in the last millennium recorded in Yongshu Reef, southern South China Sea. Palaeogeogr Palaeoclimatol Palaeoecol 210(1):89–100. doi:10.1016/j.palaeo.2004.04.002

Chapter 3
The Scientific Value of Reef-Platform Boulders for Interpreting Coastal Hazards

Abstract Coral boulders are one of the signatures of high-energy marine inundation along tropical coastlines. Data derived from boulders may include age, event frequency, inundation direction and event intensity. The latter is approached via hydrodynamic transport equations, i.e. calculation of the minimum wave energy (linked to wave height or flow velocity) required for boulder transport. However, establishing hydrodynamic models for the transport of coastal boulders involves some simplification of wave properties and transport mechanisms. Accurate dating of HEMI events through coral boulder age-dating can be achieved with traditional dating techniques (air photos, radiocarbon, uranium-series, ESR), but it also raises several challenges linked to the fundamental nature of the boulder, specifically the age of death of the corals comprising the limestone fabric.

3.1 Introduction

Understanding the timing, frequency and intensity of past high-energy marine inundation (HEMI) events is crucial for coastal planning to reduce societal vulnerability to future hazards. Unfortunately, this goal is hindered by the reality that documentary records are often hard to obtain and incomplete. Historical storm records in many places, for instance, are sparse and usually cover a period of less than 100 years (Yu et al. 2009). Longer-term written histories, i.e. spanning centennial to millennial timescales, are limited to a few regions with long periods of civilisation (Liu et al. 2001) and a tradition of careful record keeping, such as Japan and China. Where written accounts of HEMI events are short and patchy in scope, for instance in the islands of the South Pacific (Terry and Etienne 2010a), it is impossible to determine the physical vulnerability of a particular location based on historical records alone (Nott 2000). This conundrum means that geological information stored in coastal depositional sequences can provide valuable insights, if interpreted with care, on the characteristics of past HEMI events back in time beyond the documentary record, which helps in the assessment of continuing local vulnerability to coastal hazards (Richmond et al. 2011a).

Coastal boulders, including reef-platform coral boulders (RPCBs) on tropical coastlines, are one of the many signatures of HEMI events, alongside other coastal sedimentary units such as sand sheets and micro-fossil assemblages. Analysis of coastal boulders may therefore yield valuable information on prehistorical HEMI events at specific localities and this allows estimation of the potential for future marine inundation hazards. Data derived from boulders may include age, event frequency, inundation direction, spatial extent, and wave energy. In the South China Sea, for example, Yu et al. (2009) reconstructed the intensity and timing of past storms (or tsunamis) by examining the age, positions and sizes of transported coral blocks that are widely distributed on the Yongshu Reef in the Nansha area (Spratly Islands). The authors found evidence for six significant events over the last 4000 years. Similarly, from the distribution and characteristics of RPCBs at Pakarang Cape in southern Thailand, Goto et al. (2007) obtained data on inundation patterns and tsunami-wave behaviour of the 2004 Indian Ocean tsunami (IOT). The following sections consider the advantages and application of boulder studies to the field of coastal hazard research by focusing on some of the contributions that such investigations can make with interpreting the nature and characteristics of past HEMI events. The coverage pays most attention to RPCBs, but mentions boulders of other lithologies where relevant.

3.2 Coastal Sedimentology Within Marine Inundation Research

This section outlines the place of coastal boulder studies within the broader sphere of coastal sedimentology and how this relates to research on HEMI events. Discussion is therefore not restricted to coral boulders found on tropical coastlines, but considers other rock types and extra-tropical locations where helpful. For a comprehensive treatment of coastal boulders as signatures of storms in particular (generally outside the tropics), readers are directed to the review by Paris et al. (2011) and the Special Issue of *Marine Geology* dedicated to the subject (volume 283) titled *"Boulders as a signature of storms on rock coasts"*.

Early published accounts of coastal boulders being emplaced or displaced by extreme waves date back to the late nineteenth century. For example, clifftop boulder ridges in Ireland were described in an 1871 Irish Geological Survey report. Mention was made that "a block 15 × 12 × 4 ft (approximately 20 m^3) seems to have been moved 20 yards [6 m] and left on a step 10 ft [3 m] higher than its original site", which was attributed to a storm event (Kinahan et al. 1871; as reported in Williams and Hall 2004, pp. 111–112). In spite of this encouraging start, somewhat surprisingly coastal boulders have not featured as a principal focus within research on coastal sedimentology until more recent decades, when Hernandez-Avila et al. (1977) described the formation of hurricane boulder ramparts on Grand Cayman Island in the Caribbean and Bourrouilh-Le Jan and Talandier (1985) examined a 1500-ton coral block on Rangiroa Atoll in French Polynesia. Since the 1980s, however, a plethora of coastal boulder studies worldwide have been published, mostly investigating boulders as a surrogate for identifying and characterising HEMI events on coastlines.

A flavour of the impressive spread of locations where boulders have been analysed for this and related purposes can be given in an alphabetical list that includes Algeria (Maouche et al. 2009), Australia (Young et al. 1996; Nott 1997; Zhao et al. 2009; Yu et al. 2012), Bahamas (Hearty 1997), British Virgin Islands (Buckley et al. 2012) and various other Caribbean Islands (Scheffers 2002; Scheffers et al. 2005; Morton et al. 2006; Spiske et al. 2008; Pignatelli et al. 2009; Scheffers et al. 2010; Watt et al. 2010, Engel and May 2012), China (Yu et al. 2004; 2009), Fiji (Rahiman et al. 2007; Etienne and Terry 2012), France (Regnauld et al. 2010), French Polynesia (Etienne 2012), Grand Cayman (Hernandez-Avila et al. 1977; Jones and Hunter 1992), Hawaii (Noormets et al. 2002; Richmond et al. 2011b), Iceland (Etienne and Paris 2010), Italy (Mastronuzzi and Sansò 2004, Mastronuzzi et al. 2007, Scicchitano et al. 2007), Iran (Shah-hosseini et al. 2011), Ireland (Hall et al. 2010; Cox et al. 2012), Jamaica (Robinson et al. 2005), Japan (Suzuki et al. 2008; Goto et al. 2009, 2010a, 2010b, 2010c, 2011); Morocco (Mhammdi et al. 2008; Medina et al. 2011), New Zealand (Kennedy et al. 2007), Portugal (Costa et al. 2011), Puerto Rico (Taggart et al. 1993), Scotland (Hansom and Hall 2009; Scheffers et al. 2009a; Hall et al. 2010), Thailand (Goto et al. 2007) and Tonga (Frohlich et al. 2009).

Together with this remarkable upsurge in interest, studies of coastal boulders have at the same time progressed from being primarily descriptive to far more quantitative in nature. One reason for this is that a popular research aim nowadays is to better understand features of HEMI-event wave behaviour, which is crucial for assessing hazard risk on populated coastlines. An outcome of this has been the development of numerous numerical equations for estimating the wave energy required to transport coastal boulders of commonly-observed sizes (see Sect. 3.3).

While Mastronuzzi and Sansò (2004) noted that the impact of tsunamis on coasts had earlier been generally neglected compared to storms in coastal boulder research, it is true to say that the 2004 Indian Ocean tsunami (IOT), which claimed more than 230,000 lives, has done more to raise worldwide awareness of tsunami hazards than any other event. The 2004 IOT was thus a significant turning point in coastal boulder research that is now more heavily focused on tsunami than storm deposits as a result (Paris et al. 2011). (Distinguishing storm from tsunami deposits is considered in Chap. 4). Tsunami deposits containing boulders in Thailand and Indonesia laid down by the IOT have been extensively studied in the past few years (Goto et al. 2007, 2010c; Kelletat et al. 2007; Paris et al. 2010; Nandasena et al. 2011a). Similarly, the 2009 South Pacific tsunami (Etienne et al. 2011; McAdoo et al. 2011; Richmond et al. 2011a) and the 2010 Chile tsunami (Spiske and Bahlburg 2011) have also provided rare opportunities to study coastal boulders resulting from known sources of large waves. To emphasise the subset of work on RPCBs (rather than boulders of other rock types in non-reef settings), Table 3.1 highlights some of the major publications that have examined coral boulders as a tool for interpreting HEMI events. Locations of these studies are illustrated in the map in Fig. 3.1.

Table 3.1 Examples of published studies that have included an examination of coral boulders for interpreting HEMI events (in chronological order of publication)

Publication	Study area	Boulder characteristics	Cause/suggested cause
Site-specific case study focusing mainly on coral boulders			
Taggart et al. (1993)	Isla de Mona, Puerto Rico	Large boulders up to 5 m across were deposited within the last 3000 years	
Yu et al. (2004)	Nansha area (Spratly Islands), South China Sea	Boulders on reef flats	Storms likely
Goto et al. (2007)	Pakarang Cape, Thailand	Reef rock fragments up to 14 m^3	2004 Indian Ocean tsunami
Rahiman et al. (2007)	Viti Levu island, Fiji	Numerous coral boulders deposited, largest weighing 50–80 tons, volumes 20–30 m^3	1953 Suva tsunami
Suzuki et al. (2008)	Eastern Ishigaki island, Japan	Numerous massive coral boulders on the shore and in the reef moat	1771 Meiwa tsunami
Goto et al. (2009)	Kudaka island, Japan	210 boulders	Storm
Zhao et al. (2009)	Heron and Wistari reefs, GBR, Australia	Numerous (probably thousands, some a few meters across) on reef flats of two reefs	Storms
Frohlich et al. (2009)	Tongatapu island, Tonga	Seven coral limestone blocks, largest sized 15 × 11 × 9 m, located 130 m from present shoreline and 10 m above sea level	A prehistoric tsunami
Goto et al. (2010a)	Miyako-Yaeyama islands, Japan	Large, named coral boulders recorded in early documents	1771 Meiwa tsunami
Goto et al. (2010b, c)	Ishigaki island, Japan		
Paris et al. (2010)	Lhok Nga, Indonesia		2004 Indian Ocean tsunami
Terry and Etienne (2010a)	Taveuni and Vanua Levu islands, Fiji; Tupai Atoll, French Polynesia	Boulders on reef flats	Storms
Goto et al. (2011)	Okinawan Islands, Japan	Sliding and overturning of a 100-ton boulder; displacement of a 200-ton boulder	Storms
Engel and May (2012)	Bonaire, Leeward Antilles, Caribbean	Reef-platform coral boulders	Holocene tsunamis and recent storms

(continued)

Table 3.1 (continued)

Publication	Study area	Boulder characteristics	Cause/suggested cause
Yu et al. (2012)	Heron reef, Great Barrier Reef, Australia	102 transported coral boulders	Storms
Coral boulders examined/mentioned within broader investigations			
Umbgrove (1947)	Flores, Indonesia	Coral boulders of up to 10 m^3 onto beaches	1928 Paloeweh volcanic eruption; three tsunami waves of 5–10 m high
Stoddart (1969)	Rangiroa Atoll, Tuamotu Archipelago, French Polynesia	Blocks on the reef flat, largest measures 18 × 18 × 15 feet (about 5.5 × 5.5 × 4.6 m)	Storm likely
Simkin and Fiske (1983)	Anyer, Java, Indonesia	A 300-m^3 block torn from offshore reefs and deposited inland	1883 Krakatau volcanic eruption tsunami
Bourrouilh-Le Jan and Talandier (1985)	Rangiroa Atoll, Tuamotu Archipelago, French Polynesia	Largest weight 1500 tons, 15 × 10 × 5 m	
Scheffers (2002)	Leeward Antilles		Mid-Holocene storms and tsunamis
Terry (2007)	Niue, South Pacific	Many angular coral boulders weighing several tons thrown onto the Alofi Terrace 32 m above sea level	2004 Tropical Cyclone Heta
Kelletat et al. (2007)	Ko Phi Phi, Thailand	Largest weights >40 tons	2004 Indian Ocean tsunami
Scheffers (2008)	Phuket, Thailand	A 10-ton coral boulder deposited 200 m from foreshore	2004 Indian Ocean tsunami
Paris et al. (2009)	Lhok Nga, Indonesia	15 coral boulders at Lhok Nga Point; an 11-ton boulder transported 400 m inland	2004 Indian Ocean tsunami
Yu et al. (2009)	Nansha area (Spratly Islands, South China Sea) China	Boulders on reef flats	Storms likely
Omoto (2010)	Miyako Island, SW Japan	Porites boulders	tsunami

Table 3.1 (continued)

Publication	Study area	Boulder characteristics	Cause/suggested cause
Terry and Etienne (2010b)	Tetiaroa Atoll, French Polynesia	Large coral boulders uprooted, scattered hundreds of meter across reef flat	2010 Tropical Cyclone Oli
Watt et al. (2010)	Bonaire, Leeward Antilles	Coral and limestone boulders on a Pleistocene limestone platform 3–8 m above sea level	Hurricanes and tsunamis spanning 10–1000 s of years
Etienne et al. (2011)	Lhok Nga, Indonesia	>80 coral boulders	2004 Indian Ocean tsunami
Goff (2011)	Mangaia, Cook Islands	Coral boulders found 100–150 m inland; largest size $2.3 \times 2.4 \times 1.2$ m	2010 unrecorded local tsunami, possibly generated by submarine landslide
Etienne (2012)	French Polynesia	Imbricated boulders (a-axis >1 m) concentrated in reef grooves; uprooted massive coral colonies transported on the outer reef slope	2010 Tropical Cyclone Oli
Numerical model developed from transported coral boulders			
Nott (1997)	Great Barrier Reef, Australia		Likely tsunamis, possibly storm swells
Imamura et al. (2008)	Japan	Largest clast weighs 700 tons	
Nandasena et al. (2011a)	Multiple cases		

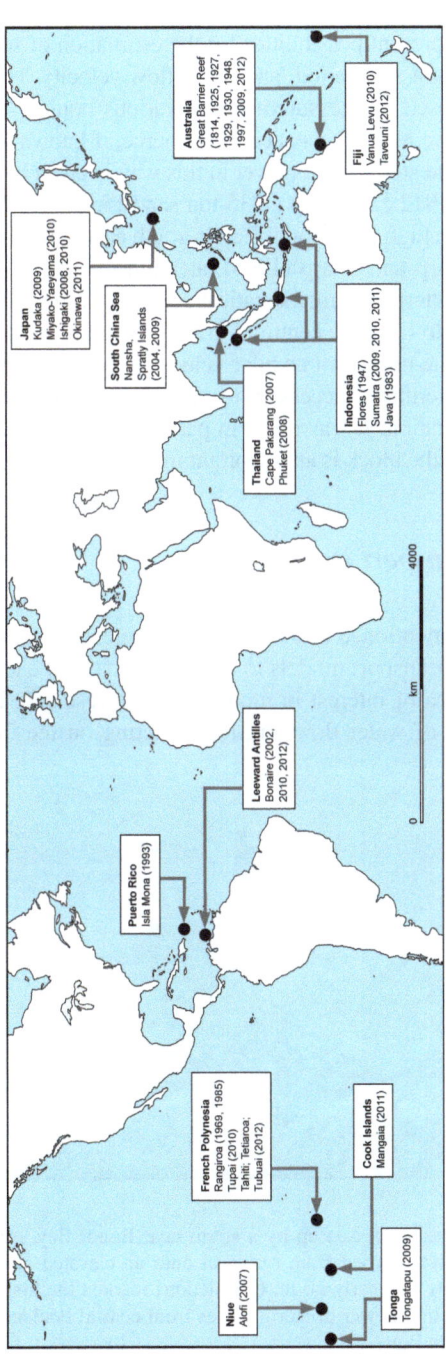

Fig. 3.1 Distribution of carbonate boulder studies on tropical and sub-tropical coasts. Dates in parentheses indicate year of observation or publication

3.3 Wave Energy Estimation

Coastal boulders that have been excavated (i.e. produced), deposited or simply moved by large waves present possibilities for the estimation of several wave parameters, such as wave height and period, and water-flow velocity. This is accomplished through calculations based on measurements of various boulder dimensions including diameter, volume, weight and distance from source, if known. One example of an early attempt to calculate storm wave power in this way is given by Sussmilch (1912), where a storm in July 1912 carried up a 235-ton sandstone block from sea level onto a platform 10 feet (3 m) high near Bondi Bay in south east Australia (Fig. 3.2). Citing the same example, Sharp and Sharp (1995) later illustrated how a 'water hammer' approach provides an alternative mathematical solution to the problem. The authors used energy principles to raise the centre of gravity of the block by a known amount and thereby calculate the minimum angular velocity required. A wave impact velocity of 16.4 m/s was determined necessary to raise the boulder, assuming lifting was achieved by angular rotation. It was noted in particular how the elastic behaviour of water confined beneath the block is an important influence affecting its (in)stability.

3.3.1 Boulder Transport Equations

Following the studies mentioned above, the development of various hydrodynamic equations and boulder transport models to estimate wave and energy conditions has grown into a major area of interest in modern coastal research. Hydrodynamics is defined as the science of water flow and forces acting on the flow (Hearn 2011).

Fig. 3.2 A giant sandstone block cast up by a storm near Bondi Bay, Australia in July 1912. The joint-bounded block was carried from sea level onto an elevated platform over 3 m high and transported horizontally by nearly 50 m. Calculations marked the first documented attempt (1912) to determine the lifting power of storm waves from coastal boulder displacement. Photo from Sussmilch (1912, plate IV)

Among existing boulder hydrodynamic equations, the most popular are those of Nott (1997, 2003) cited 94 and 88 times respectively.[1] Nott's equations allow calculation of the minimum heights of storm and tsunami waves necessary for initiating boulder movement in three different pre-transport settings: subaerial, submerged and joint-bounded settings. Parameters within the equations include boulder and water density, boulder dimensions, and coefficients of mass, drag, lift, inertia and gravity (Box 2). The underlying principle is that a boulder is only set in motion once wave energy overcomes the forces resisting boulder movement. For boulders located on subaerial shore platforms, the forces of inertia, drag and lift are experienced; for submerged boulders, inertia is absent owing to the buoyant support of water. For joint-bounded settings, a lift force is required to pluck a boulder out of its 'trap', hence greater wave height is required to initiate transport (Nott 2003).

Box 2
Hydrodynamic Equations for Coastal Boulder Transport (Nott 2003)

Scenario	Storm	Tsunami
Submerged boulders:	$H_s \geq \frac{(\rho_s - \rho_w/\rho_w)\,2a}{C_d(ac/b^2)+C_l}$	$H_t \geq \frac{0.25\,(\rho_s - \rho_w/\rho_w)\,2a}{C_d\,(ac/b^2)+C_l}$
Subaerial boulders:	$H_s \geq \frac{(\rho_s - \rho_w/\rho_w)\,[(2a-4C_m(a/b)(\ddot{u}/g))]}{C_d(ac/b^2)+C_l}$	$H_t \geq \frac{0.25\,(\rho_s - \rho_w/\rho_w)\,[(2a-C_m(a/b)\,(\ddot{u}/g))]}{C_d\,(ac/b^2)+C_l}$
Joint-bounded boulders:	$H_s \geq \frac{(\rho_s - \rho_w/\rho_w)\,a}{C_l}$	$H_t \geq \frac{0.25\,(\rho_s - \rho_w/\rho_w)\,a}{C_l}$

Parameters:

H_s, H_t	height of the storm or tsunami wave at breaking point (m)
a, b, c	long, intermediate and short axes of boulder (m)
ρ_s	density of boulder (tons/m^3 or g/cm^3)
ρ_w	water density ($= 1.02$ g/ml for sea water)
C_d	drag coefficient[a]
C_m	coefficient of mass ($= 2$)
C_l	lift coefficient ($= 0.178$)
$\ddot{u}$	instantaneous flow acceleration ($= 1$ m/s^2)
g	gravitational acceleration ($= 9.81$ m/s^2)

[a]Nott (2003) and Nott and Bryant (2003) explained that their choice of C_d values was based on the experimental work carried out by Noji et al. (1985), in which a cube was used for determining C_d as many wave-transported boulders present similar shapes. C_d was seen to vary substantially with time during the passage of waves, from approximately 1.5 to 5. Different C_d values were therefore adopted according to the *water depth : wave height* ratio in a particular coastal setting. In Nott (2003) C_d values of 2 and 1.5 were adopted for submerged and subaerial boulders respectively, and $C_d = 3$ in Nott and Bryant (2003).

---

[1] *Source* Scopus, July 2012.

Although Nott's equations have been employed by several investigators for calculating wave energy (e.g. Mastronuzzi and Sansò 2004; Scicchitano et al. 2007; Spiske et al. 2008; Etienne and Paris 2010), more recent work argues that improvements are necessary to overcome several ambiguities that have since been identified in the original equations. By applying the equations to situations where wave height is known, it has been recognised that equations may overestimate the required wave height (or underestimate the power of waves) to move boulders (Mastronuzzi and Sansò 2004; Switzer and Burston 2010; Paris et al. 2010). In addition, Nott's simple mathematical assumption of tsunami waves being four times more capable of shifting boulders than storm waves of the same wave height (Nott 2003) has been criticised by Morton et al. (2006). Presumptions of overturning (rolling) as the principal mode of transport have also been challenged, because some larger and irregularly-shaped boulders were found to have moved by sliding (Noormets et al. 2004; Morton et al. 2006).

Consequently, in fresh work by Nandasena et al. (2011a, b), Nott's original equations were reassessed by applying them to four case studies, and subsequently revised for greater accuracy. The equations were adjusted by (1) rearranging lift area for both subaerial and submerged settings, (2) removing the inappropriate use of inertial force for subaerial boulders and (3) balancing forces in the direction of lifting for joint-bounded boulders (Box 3). In the enhanced hydrodynamic equations, it is possible to predict flow velocities required to initialise boulder displacement by alternative modes of transport such as sliding and saltation/lifting. This addresses the earlier concerns raised by various authors (Williams and Hall 2004; Noormets et al. 2004).

Box 3
Revised Hydrodynamic Equations (Nandasena et al. 2011b)

Initial transport mode	Subaerial boulders	Submerged boulders	Joint-bounded boulders
Sliding:	$u^2 \geq \dfrac{2(\rho_s/\rho_w - 1)\, gc\, (\mu_s\cos\theta + \sin\theta)}{C_d(c/b) + \mu_s C_l}$		NA
Rolling or overturning:	$u^2 \geq \dfrac{2(\rho_s/\rho_w - 1)\, gc\, (\cos\theta + (c/b)\sin\theta)}{C_d(c^2/b^2) + C_l}$		NA
Saltation or lifting:	$u^2 \geq \dfrac{2(\rho_s/\rho_w - 1)\, gc\cos\theta}{C_l}$		$u^2 \geq \dfrac{2(\rho_s/\rho_w - 1)\, gc\, (\cos\theta + \mu_s\sin\theta)}{C_l}$

Parameters:

u	flow velocity (m/s)
b, c	intermediate and short axes of boulder (m)
ρ_s	density of boulder (tons/m^3 or g/cm^3)
ρ_w	water density ($=1.02$ g/ml for sea water)
C_d	drag coefficient ($=1.95$)
C_l	lift coefficient ($=0.178$)
θ	angle of bed slope at pre-transport location ($°$)
μ_s	coefficient of static friction ($=0.7$)
g	gravitational acceleration ($=9.81$ m/s^2)

In many situations, a reef-platform boulder field comprises numerous individual boulders for which the original mode of transport is unknown. In such cases a 'transport histogram' can be constructed by substituting boulder measurements into the equations of Nandasena et al. (2011b) and plotting the results for minimum water flow velocity needed to initiate boulder movement according to all possible modes of transport (e.g. Fig. 3.3). Such transport histograms are useful as they permit both comparison of results 1. between various sites of investigation worldwide and 2. between the energy characteristics of ancient and recent HEMI events at a specific location, if separate clusters of boulders delivered by more than one event can be distinguished.

Another equation has been proposed by Frohlich et al. (2011) for determining the velocity of HEMI-induced currents using a physics-based analysis of classical hydrodynamical concepts (Box 4). Noting that the drag force exerted on a boulder is approximately equivalent to its buoyant weight, the equation can be used to determine the current velocity needed to initiate motion. The authors also suggested a 'rule of thumb' for rapid estimation in the field, where "the height of a wave necessary to displace a boulder of given density and height is approximately proportional to the product of the boulder's relative density and its linear dimension" (Frohlich et al. 2011, p. 682). However, owing to the highly non-linear interaction between waves and shorelines, and the critical dependence of wave energy on the nearshore bathymetry, computer modelling has been advocated as essential to improve the accuracy of results (Mader 2004; Frohlich et al. 2011).

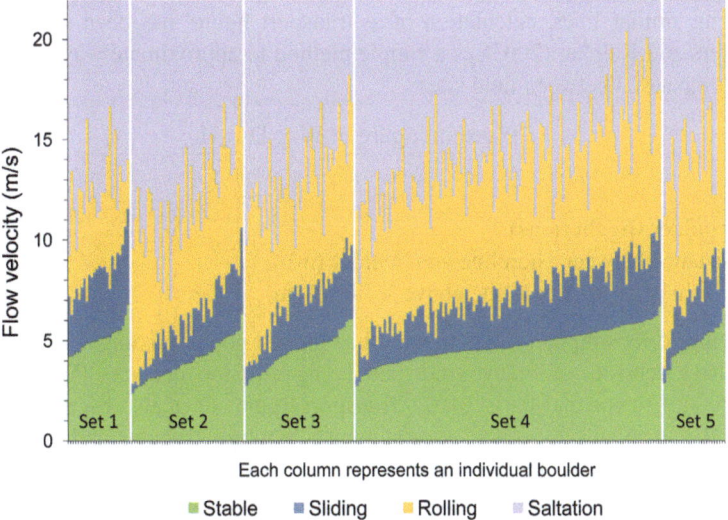

Fig. 3.3 A transport histogram for reef-platform coral boulders measured on Makemo Atoll in French Polynesia (A. Lau, unpublished data). The vertical axis shows the minimum water flow velocity required to initiate boulder movement according to three types of transport mode. Transport equations by Nandasena et al. (2011b) were used to calculate velocities (Box 3). Sets one to five represent different sites of investigation along a 15 km stretch of the northern coastline of the atoll. The horizontal axis plots individual boulders which are ranked from smallest to largest within each set

---

Box 4
Boulder Displacement: Equations for Estimating Minimum Current Velocity and Wave Height (Frohlich et al. 2011)

Minimum current velocity for boulder displacement:

$$V > \sqrt{\frac{2g}{C_d}\left(\frac{\rho_b}{\rho_w} - 1\right)h_b}$$

Minimum wave height for boulder displacement (described as "a very crude 'rule of thumb' "):

$$h_w > \frac{\rho_b}{\rho_w}h_b$$

Parameters:

V	Velocity
g	Gravitational acceleration
ρ_b	Density of boulder
ρ_w	Water density
C_d	Drag coefficient
h_b	Boulder dimension
h_w	Wave height

Along similar lines, calculation of a 'transport figure' has been suggested by Scheffers and Kelletat (2003) as a simple method to approximate wave energy for easy comparison between field sites:

$$\text{Transport figure} = W \times D \times H$$

where:

W = boulder weight (tons)
D = distance between shoreline and boulder (m)
H = height of depositional site above sea level (m)

Owing to its simplicity, this method has proved popular, being widely applied to coastal deposits of many kinds, including cliff-top deposits (Williams and Hall 2004; Etienne and Paris 2010). Transport figures have also been used, along with other available evidence, to distinguish between tsunami and storm boulder deposits, since the former supposedly exhibits far greater values than the latter. However, this technique appears unsuited to assessing the strength of HEMI events from reef-platform coral boulders, because the reefs on which RPCBs rest are generally at or near sea level. Thus, the value of H (height above sea level) approximates to zero, leading to an output of zero from the equation (multiplication product = 0), regardless of boulder size or distance transported.

In addition to the hydrodynamic equations presented above, alternative approaches have also been devised to investigate boulder transport by HEMI-event

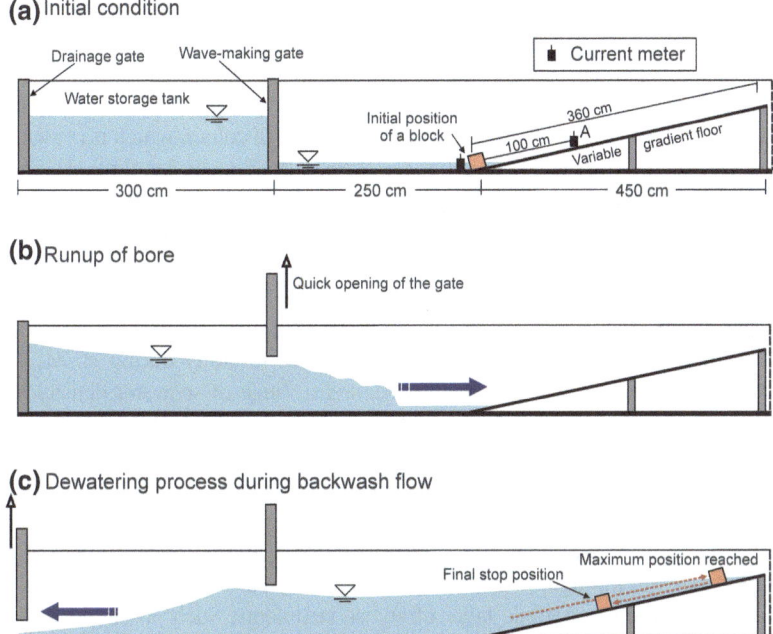

Fig. 3.4 The hydraulic experiment carried out by Imamura et al. (2008) for studying boulder transport by tsunami waves. Carbonate and silicate rock samples were used to simulate boulders in the experiment. Water current velocity was measured at point *A*. Observations determined that the rocks move by sliding, rolling and saltation. On a slope, the rock is first pushed uphill and then comes to rest at a lower position, as shown in (c). (Redrawn from Imamura et al. 2008, p. 3)

waves. By conducting hydraulic experiments in a water tank, Imamura et al. (2008) developed a numerical model to estimate hydraulic values of tsunamis from the movement of simulated boulders (rock fragments) of different lithologies (Fig. 3.4). Hydraulics is the science of the behaviour of moving fluids under pressure. The experiments revealed that clast movement is achieved by sliding, rolling or saltation, thereby illuminating the various possible modes of transport. The researchers suggested that the frictional coefficient should decrease if a boulder rolls or saltates, as contact time with the ground is reduced. This implies that boulders rolling or saltating should travel longer distances compared to sliding under the same energy conditions (Imamura et al. 2008).

3.3.2 Assumptions and Difficulties

Establishing hydrodynamic models for coastal boulders inevitably involves some simplification of wave properties and transport mechanisms. This means that any results stemming from such calculations deserve to be treated with a healthy degree of caution. For instance, clear water conditions are often assumed and represented by a seawater density of 1020 kg/m³. However, highly elevated turbidity is an observed feature of sea conditions during HEMI events, especially tsunamis (Richmond et al. 2011a; Etienne et al. 2011). Eyewitnesses on the island of Upolu in Samoa, for example, described the 2009 South Pacific tsunami as black in colour, which is evidence for the waves carrying significant quantities of suspended sediment (Dominey-Howes and Thaman 2009). Similarly, on Sumatra, accounts by local people at Lhok Nga after the 2004 IOT confirmed that the tsunami waves were dark because of their high sediment content (Lavigne et al. 2006). Large amounts of organic matter as well as man-made debris may also increase the water density of tsunami waves. This raises the chances for debris impact with obstructions, making the waves more erosive (Richmond et al. 2011a). The effects on boulder transport of elevated turbidity during HEMI events should therefore not be neglected and accepting a range of seawater density values would be a better option. Other branches of sedimentology may provide more realistic values for this parameter. A submarine debris flow triggered by an undersea landslide has a density of 1250 kg/m³, while a turbidity current can reach 1800 kg/m³ (Talling et al. 2007). Along the same lines, it may be the case that a semi-fluidized seabed of gravel and cobble material during a high energy event assists the movement of boulder-sized material, through inter-clast collision and by partially supporting the weight of boulders on top of a mobile layer of coarse sediments. Such possibilities deserve investigation, in order to determine what further developments in existing transport equations are required to render them more representative of natural situations.

Another issue recently pointed out by Weiss (2012) is that bed roughness is a major influence on incipient boulder motion. Currently, this element is absent in existing models, but could be incorporated to improve their precision. By quantifying the effect of bed roughness on an assumed circular boulder, the author demonstrated that a boulder resting on a bed with roughness greater than about 30 % of its radius is prevented from moving (Fig. 3.5). This indicates that bed roughness of the pre-transport location of coastal boulders must be surveyed in the field in addition to boulder dimensions, although it is recognised that this may be

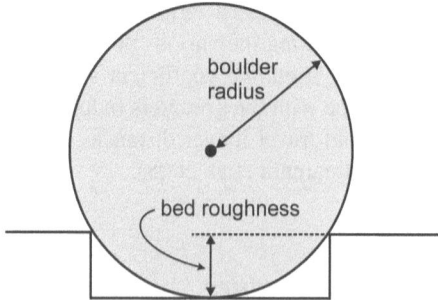

Fig. 3.5 Simplification of the influence of bed roughness on the stability of a boulder with circular geometry exposed to constant forces, as presented by Weiss (2012). Boulder movement by storm or tsunami waves will be hindered if the bed roughness exceeds approximately 30 % of the boulder radius

problematic if the surface roughness was created by smaller boulders and cobbles that were also redistributed by the same HEMI event. Overall, Weiss (2012 p. 33) concluded that "boulder transport might not be fully understood until three-dimensional flow simulation around boulders with arbitrary geometries is carried out to study the three dimensional stress distribution".

3.4 Inundation Direction

The orientations of coastal boulders have been examined to infer wave direction during HEMI events (Hall et al. 2006; Paris et al. 2010; Etienne et al. 2011). Individual clasts tend to be organised in a way that is most stable against flow. There are two components to this arrangement. First, clasts orient themselves in a pattern known as imbrication, where they dip towards the flow source. Second, for clasts with elongated shapes, the longest axis usually lies perpendicular (normal) to the water flow (Nichols 1999) (Fig. 3.6). Thus, for coastal boulders too big to be oriented by waves under normal conditions, the average long-axis orientation and dip direction for groups of boulders provide indicators for the direction of HEMI-event wave approach, assuming that the energy of backwash was weaker than the swash and hence did not rearrange the clasts. Since the inundation direction of each HEMI event at a certain locality is likely to differ, coastal boulders emplaced by subsequent events at that location should be imbricated and orientated in a variety of directions. Consequently, imbrication and orientation data measured from sets of boulders can help in identifying and differentiating between a number of discrete past inundation events that have affected a certain section of coastline.

Some researchers have extended these ideas further to suggest that clast orientations may also help to indicate transport modes. Assuming that extreme-wave flow directions are approximately normal to the shore, the principle is that clasts oriented parallel to the shoreline (normal to flow) indicate they moved by rolling or sliding, as these transport modes require relatively lower energy. In contrast, when clast orientations are normal to the shore (parallel to flow), this implies they were transported by more turbulent flow in suspension or by saltation, without significant rolling (Williams and Hall 2004; Watt et al. 2010).

3.5 Boulder Mapping

In coastal hazard studies, marine sediment mapping is often carried out to define inundation limits of modern events (e.g. Paris et al. 2009; Richmond et al. 2011a). Boulder location can be achieved by field measurement (see Chap. 4) or by image analysis. The latter includes interpretation of vertical aerial photography, although air photo surveys are seldom achieved soon after a HEMI event. Oblique aerial pictures, often provided by national or international rescue teams, offer general views

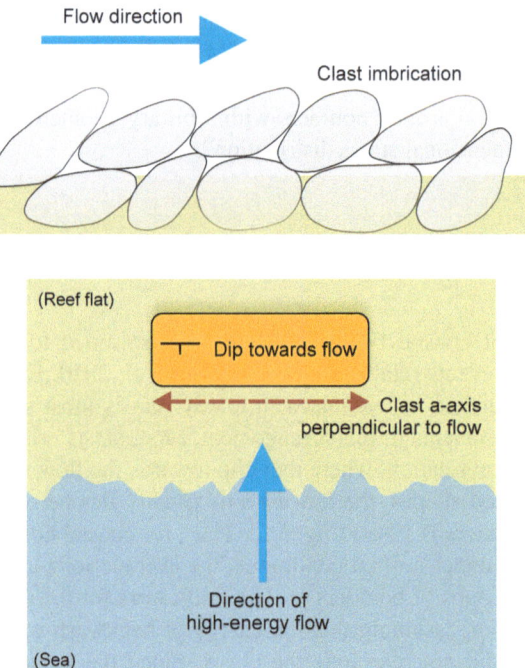

Fig. 3.6 Imbrication of clasts within coarse deposits (Redrawn from Nichols 1999). *Above*: cross-section view. Clasts imbricate themselves in an arrangement that is most stable against the flow of water and come to rest dipping upstream. *Below*: aerial cartoon of a reef coast to illustrate clast orientation in relation to water flow (not to scale). The rounded rectangle represents an elongated coastal boulder moved during high energy wave conditions. To stay stable in the flow of water, the boulder dips towards the flow source and is oriented with the longest axis (a-axis) perpendicular to the flow. By measuring the dip direction and orientation of coastal boulders, inundation direction can therefore be inferred

of sediment fields. More rarely, video footage during an event can be used for scientific purposes (e.g. Hall 2011; Tappin et al. 2012), but is not normally satisfactory for accurate boulder positioning. However, the use of drones or remotely-controlled helicopters shows promise as a way to improve boulder field mapping in the near future.

Moreover, very high resolution (VHR) remote sensing is continuing to develop. Remote sensing (from satellites or airborne sensors) has been widely used for characterizing coastal environments: geomorphological mapping, vegetation mapping, ecological studies, coastal zone management, etc., although no study has yet focussed on the use of satellite imagery specifically for coastal boulder mapping. This can be explained by the pixel resolution which was not previously in accordance with boulder size. However, VHR spatial imagery at sub-metre resolution is now available and the level of spectral mixing encountered in lower resolution imagery is becoming more manageable (Table 3.2). Multispectral images released since 2009 by the Digital Globe® WorldView-2 satellite offer a 46 cm resolution in panchromatic mode and include a new 'purple' band (400–450 nm) optimized for coastal studies

Table 3.2 Opportunities for coastal boulder observation with very high resolution satellite imagery

Satellite	Company	Launch date	Number of bands	Resolution at nadir, panchromatic mode (m)	Resolution at nadir, multi-spectral mode (m)
IKONOS	GeoEye (US)	24 Sept 1999	5	0.82	3.28
QuickBird-2	DigitalGlobe (US)	18 Oct 2001	5	0.61/0.65[a]	2.44/2.62[a]
WorldView-1	DigitalGlobe (US)	18 Sept 2007	1	0.41[b]	n.a
GeoEye-1	GeoEye (US)	6 Sept 2008	5	0.41[b]	1.65
WorldView-2	DigitalGlobe (US)	8 Oct 2009	9	0.46[b]	1.84
Pleiades-1A	Spot Image (Fr)	17 Dec 2011	5	0.70[c]	2.80
To be launched					
GeoEye-2	GeoEye (US)	Expected 2013	5	0.34	1.36

[a] depending on the altitude of measurement (450/482 km respectively)
[b] imagery re-sampled to 0.5 m for all customers not explicitly granted a waiver by the U.S. Government
[c] interpolated to 50 cm

(vegetation and coral mapping, bathymetry extraction; Fig. 3.7). Also, the GeoEye-2 satellite is due to be launched in 2013 and will release images with a greater resolution still (34 cm in panchromatic mode). Remotely sensed imagery allows the generation of thematic maps based on image classification. Classification relies on differences in spectral reflectance (R) between classes. The spectral signature of boulder deposits might then be used to isolate population classes. It is presumed that, for the same lithology, variations in R reflect different stages of weathering, hence different ages. Boulders with close R values will be considered as elements within the same population, i.e. belonging to the same HEMI event. This innovative methodological approach combines both traditional pixel-oriented (spectral) classification and object-oriented classification. The use of remote sensing for coastal boulder mapping is therefore anticipated to increase significantly over coming years.

3.6 Dating Prehistorical HEMI Events

3.6.1 Boulder Age-Dating

Two principal methods exist for approximating the age of past HEMI events on affected coastlines from the analysis of coastal boulders. One possibility is to compare successive sets of photographs, aerial photos or high-resolution satellite images for a certain location (Goto et al. 2009, 2010b, 2011). Although not normally providing an exact age, examination of time-series imagery may effectively bracket a specific period of time within which the emplacement and/or movement of boulders occurred.

Alternatively, various dating techniques are available. Lichenometry has been employed by Hall et al. (2006) for relative-age dating of cliff-top storm deposits

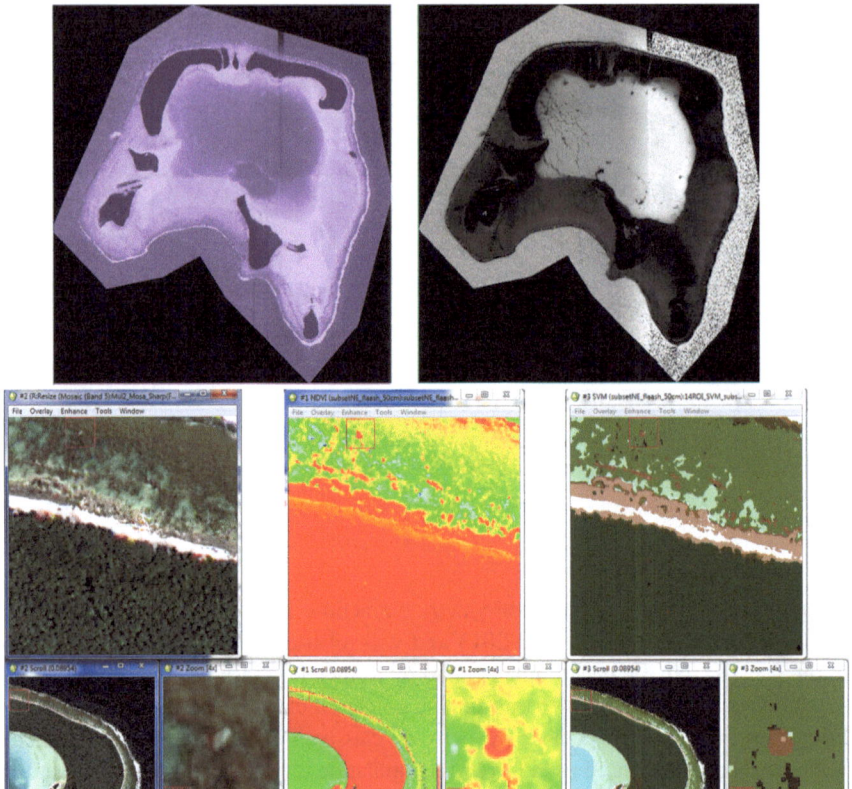

Fig. 3.7 Analysis of a WorldView-2 image of Tetiaroa Atoll in French Polynesia. *Upper Left*: raw 'coastal' band (400–450 nm). *Upper right*: bathymetry extraction. *Below*: coastal boulder identification via SVM classification (*SVM* Support Vector Machine, now a standard form of image analysis in remote sensing, where automated object recognition is possible after 'training' the system the recognise objects of interest). Courtesy of A. Collin, unpublished data

in Scotland. The technique involves estimating the residence time of boulders by observing the amount of lichen cover. Lichen growth rates are species dependent, but in general a lichen-covered boulder indicates a minimum residence time of several decades. Hall et al. (2008) estimated that it takes at least 70 years for the black tar lichen *Verrucaria maura* to colonise more than 50 % of boulder surfaces in the Shetland Islands of northern Scotland.

For much older prehistorical events, sediment age-dating is more appropriate. With reef-platform coral boulders comprising carbonate material, dating of the organic framework (i.e. fossil corals) is possible. If it is assumed that living corals died at the time a fragment of the reef structure was broken away to form a new RPCB by a HEMI event, then the age of coral mortality approximates the timing of the event itself (Yu et al. 2009). Preferably the youngest part of a coral boulder needs to be sampled for analysis (top stratigraphic layer), if this can be determined from the growth direction of fossil corals within a boulder's fabric.

Fig. 3.8 A volcanic boulder encrusted with living marine molluscs and barnacles (foreground), on the intertidal reef platform near Lavena village on the eastern coast of Taveuni Island in Fiji. If similar coastal boulders were found deposited inland, then the shells of these molluscs provide organic material for dating, giving the approximate timing of the HEMI event that removed the boulder from the marine environment. (Photo by J. Terry, July 2010)

For non-carbonate boulder lithologies, alternative materials are needed for dating. Suitable materials may be provided by marine organisms inhabiting a boulder's surface. Organisms include oysters, limpets, mussels, chitons and other marine molluscs (Fig. 3.8); barnacles; and tube-building marine worms such as the *Serpulidae* family (serpulids). It is presumed that removal of the boulder from the marine environment by a HEMI event caused the mortality of the encrusting organisms. The mortality age of the youngest organic material then approximates the date of the HEMI event (Nott 2004; Scicchitano et al. 2007; Maouche et al. 2009).

Laboratory methods based on the analysis of radiocarbon (carbon-14), uranium-series (U-series) and electron spin resonance (ESR) are available for dating carbonate materials. Radiocarbon dating has been commonly used on coastal deposits (e.g. Nott 1997; Goto et al. 2010d). Errors in radiocarbon age may be in the order of $\pm$ 60–70 years over Holocene timescales (Hayne and Chappell 2001), although the method is less suited to 'recent' boulders formed within the last 350 years or so because of calibration difficulties. However, so-called 'wiggle-matching' is a relatively new technique that attempts to match short-term fluctuations in the radiocarbon calibration curve to a sequence of ^{14}C dates (e.g. from a coral core). If certain conditions are met, errors can be reduced to a minimum, thereby yielding much higher-precision dates than was previously possible (Walker 2005; Gale 2009).

Uranium-series dating measures the ratio of ^{238}U : ^{234}U : ^{230}Th (uranium-238 : uranium-234 : thorium-230) to obtain the age of a carbonate sample (Zhao et al. 2001, 2009). In coral growth, a small amount of uranium from sea water is incorporated into the CaCO$_3$ mineral (calcium carbonate), whereas thorium is not present owing to its absence in sea water (Emiliani 1992). Knowing the half-lives of ^{234}U and ^{230}Th to be 244,600 and 75,380 years respectively, and the initial ^{234}U : ^{238}U ratio determined from living coral, the age of dead coral can be measured (Zhao et al. 2001). Using the modern TIMS (Thermal Ionization Mass Spectrometry) U-series method, coral boulders may be dated with remarkable precision, to within 1–5 years accuracy (or 1–2 % uncertainty) for fossil corals younger than 1000 years old (Cobb et al. 2003; Yu et al. 2009; Frohlich et al. 2009).

The electron spin resonance (ESR) spectroscopy method is based on radiation exposure. Through bombardment by natural radiation, electrons and free radicals are 'trapped' and accumulated in crystalline minerals. By measuring the amount of trapped electronic charge in minerals like calcite and aragonite, the age of coralline sediments can thus be measured. The error of ESR coral dating ranges between 5–8 %, but the upper limit of measurement above 500,000 years extends beyond the range of radiocarbon and U-series dating (Schellmann et al. 2008). Table 3.3 provides examples of published work where various types of boulder dating methods have been used.

3.6.2 Obstacles to Accurate Dating

Clearly, the analysis of time-series aerial photographs to identify the appearance of new coastal boulders, or the displacement of pre-existing ones, is limited to relatively modern HEMI events and is only applicable on coastlines where the necessary imagery is available. In contrast, sediment age-dating methods offer possibilities for estimating the age of much older events. However, there are several sampling issues that require special attention with particular reference to dating RPCBs. A crucial priority is to sample the youngest face of coral boulders, so as to determine the mortality age of the most-recently growing corals. This is because corals grow at relatively slow rates of mm to cm per year. Corals comprising opposite faces of a large RCPB over a metre in diameter could potentially be decades or hundreds of years apart in age. Unfortunately, identifying the youngest face may not be a simple task when the coral structure has been altered by surface erosion or weathering. Also, boulders are often overturned during transportation (Nott 1997), which may hinder the identification of the youngest part of the boulder if a 'way-up' indicator is lacking. Compounding the problem is that coral boulders cast up onto intertidal reef platforms will then suffer denudation by weathering, wave action and bioerosion processes post-deposition (Fig. 3.9). This means that the original (youngest) boulder face is gradually worn away over time. A related concern is that boulder surfaces attacked by marine boring organisms must be avoided, since the incorporation of any recent organic material introduces serious errors in dating.

Another difficulty is with the assumption that the corals comprising a RPCB were killed by the HEMI event that originally quarried the boulder from the

Table 3.3 Coastal boulder dating to assess the age of past HEMI events: some published examples

Dating method	Authors	Study site	Dated material	Age and notes
For coral boulders				
TIMS uranium-series	Yu et al. (2004)	Yongshu reef, Nansha area, Spratly Islands, South China sea	Coral boulders and in situ reef flat corals	Six major storm events: 1064 ± 30 AD; 1210 ± 5–1201 ± 4 AD; 1336 ± 9 AD; 1443 ± 9 AD; 1685 ± 8–1680 ± 6 AD; 1872 ± 15 AD
Electron spin resonance	Scheffers et al. (2009b)	Caribbean	Coral boulders, coral rubble ridges and other coral debris	Eleuthera, Bahamas: 3000 years BP; Bonaire, Lesser Antilles: 1300 years BP
Radiocarbon	Nott (1997)	Great Barrier Reef, Australia	Coral boulders on reef flats	Mean age of 481 ± 27 years BP
For coastal boulders of non-carbonate lithologies				
Radiocarbon	Scicchitano et al. (2007)	Sicily, Italy	Encrusting organisms: serpulids (worms) and balanids (barnacles)	Three clusters of ^{14}C ages, suggesting at least 3 large tsunamis probably triggered by earthquakes in 1169, 1693 and 1908 AD
Lichenometry	Hall et al. (2006)	Shetland Islands, Scotland; Inish Mór, Aran Islands, Ireland	Black tar lichen (*Verrucaria maura*) coverage on sandstone boulders	Rocks with limited *V. maura* cover are exposed for <70 years. Isolated lichen-covered rocks on Inish Mór tentatively linked to a major storm in 1839
Optically stimulated luminescence	Hall et al. (2006)	Shetland Islands, Scotland	Intercalated sands	Oldest about 800 AD

Fig. 3.9 *Above*: the chief of Navola village on the 'Coral Coast' (South coast) of Viti Levu Island in Fiji stands in front of an irregular block of coral limestone (reef rock) perched just a few metres from the edge of the fringing reef. The block was emplaced on the reef flat before living memory. The highly-pitted surface is a result of solution weathering, marine abrasion and bioerosion. Fossil corals are not clearly visible, so hindering recognition of the original direction of coral growth. *Below*: marine boring organisms must be excluded from carbonate samples that are collected for dating purposes. (Photos by J. Terry, February 2010 (*above*) and June 2012 (*below*))

living reef framework. This assumption might be challenged because the alternative possibility exists that coral mortality occurred before the HEMI event (Goto et al. 2007). This might happen if a RPCB is excavated from older parts of a reef structure. In this case, boulder age then provides only a maximum age limit for the event in question (Yu et al. 2009). Finally, pre-existing boulders on reef flats are often remobilised in subsequent HEMI events (see Chap. 4). In such circumstances, the timing of subsequent events cannot be deduced from sediment-age dating as described earlier.

References

Bourrouilh-Le Jan FG, Talandier J (1985) Sédimentation et fracturation de haute énergie en milieu récifal: tsunamis, ouragans et cyclones et leurs effets sur la sédimentologie de la géomorphologie d'un atoll: motu et hoa, à Rangiroa, Tuamotu, SE Pacifique. Mar Geol 67:263–333. doi:10.1016/0025-3227(85)90095-7

Buckley ML, Wei Y, Jaffe BE, Watt SG (2012) Inverse modeling of velocities and inferred cause of overwash that emplaced inland fields of boulders at Anegada, British Virgin Islands. Nat Hazards 63(1):133–149. doi:10.1007/s11069-011-9725-8

Cobb KM, Charles CD, Cheng H, Kastner M, Edwards RL (2003) U/Th-dating living and young fossil corals from the central tropical Pacific. Earth Planet Sci Lett 210(1–2):91–103. doi:10.1007/s11069-011-9725-8

Costa PJM, Andrade C, Freitas MC, Oliveira MA, da Silva CM, Omira R, Taborda R, Baptista MA, Dawson AG (2011) Boulder deposition during major tsunami events. Earth Surf Proc Land 36:2054–2068

Cox R, Zentner DB, Kirchner BJ, Cook MS (2012) Boulder Ridges on the Aran Islands (Ireland): recent movements caused by storm waves, not tsunamis. J Geol 120:249–272. doi:10.1086/664787

Dominey-Howes D, Thaman R (2009) UNESCO-IOC International Tsunami Survey Team Samoa (ITST Samoa). Interim report of field survey 14–21st Oct 2009, p. 172. Australian Tsunami Research Centre, Sydney (Unpublished Report)

Emiliani C (1992) Planet Earth: cosmology, geology, and the evolution of life and environment. Cambridge University Press, Cambridge

Engel M, May SM (2012) Bonaire's boulder fields revisited: evidence for Holocene tsunami impact on the Leeward Antilles. Quatern Sci Rev 54:126–141. doi:10.1016/j.quascirev.2011.12.011

Etienne S (2012) Marine inundation hazards in French Polynesia: geomorphic impacts of Tropical Cyclone Oli in February 2010. Geol Soc Lond Spec Publ 361:21–39. doi:10.1144/SP361.4

Etienne S, Paris R (2010) Boulder accumulations related to storms on the south coast of the Reykjanes Peninsula (Iceland). Geomorphology 114:55–70. doi:10.1016/j.geomorph.2009.02.008

Etienne S, Buckley M, Paris R, Nandasena AK, Clark K, Chagué-Goff C, Goff J, Richmond B (2011) The use of boulders for characterizing past tsunamis: lessons from the 2004 Indian Ocean and 2009 South Pacific tsunamis. Earth Sci Rev 107:75–90. doi:10.1016/j.earscirev.2010.12.006

Etienne S, Terry JP (2012) Coral boulders, gravel tongues and sand sheets: features of coastal accretion and sediment nourishment by Cyclone Tomas (March 2010) on Taveuni Island, Fiji. Geomorphology 175–176:54–65

Frohlich C, Hornbach MJ, Taylor FW, Shen C, Moala A, Morton AE, Kruger J (2009) Huge erratic boulders in Tonga deposited by a prehistoric tsunami. Geology 37:131–134. doi:10.1130/G25277A.1

Frohlich C, Hornbach MJ, Taylor FW (2011) Megablocks. In: Hopley D (ed) Encyclopedia of modern coral reefs: structure, form and process. Springer Netherlands, Dordrecht, pp 679–683

Gale SJ (2009) Dating the recent past. Quat Geochronol 4:374–377. doi:10.1016/j.quageo.2009.05.011

Goff J (2011) Evidence of a previously unrecorded local tsunami, 13 April 2010, Cook Islands: implications for Pacific Island countries. Nat Hazards Earth Syst Sci 11:1371–1379. doi:10.5194/nhess-11-1371-2011

Goto K, Chavanich SA, Imamura F, Kunthasap P, Matsui T, Minoura K, Sugawara D, Yanagisawa H (2007) Distribution, origin and transport process of boulders deposited by the 2004 Indian Ocean tsunami at Pakarang Cape, Thailand. Sed Geol 202(4):821–837. doi:10.1016/j.sedgeo.2007.09.004

Goto K, Okada K, Imamura F (2009) Characteristics and hydrodynamics of boulders transported by storm waves at Kudaka Island, Japan. Mar Geol 262:14–24. doi:10.1016/j.margeo.2009.03.001

Goto K, Kawana T, Imamura F (2010a) Historical and geological evidence of boulders deposited by tsunamis, southern Ryukyu Islands. Jpn Earth-Sci Rev 102(1–2):77–99. doi:10.1016/j.earscirev.2010.06.005

Goto K, Miyagi K, Kawamata H, Imamura F (2010b) Discrimination of boulders deposited by tsunamis and storm waves at Ishigaki Island, Japan. Mar Geol 269:34–45. doi:10.1016/j.margeo.2009.12.004

Goto K, Shinozaki T, Minoura K, Okada K, Sugawara D, Imamura F (2010c) Distribution of boulders at Miyara Bay of Ishigaki Island, Japan: a flow characteristic indicator of tsunami and storm waves. Island Arc 19(3):412–426. doi:10.1111/j.1440-1738.2010.00721.x

Goto K, Okada K, Imamura F (2010d) Numerical analysis of boulder transport by the 2004 Indian Ocean tsunami at Pakarang Cape, Thailand. Mar Geol 268:97–105. doi:10.1016/j.margeo.2009.10.023

Goto K, Miyagi K, Kawana T, Takahashi J, Imamura F (2011) Emplacement and movement of boulders by known storm waves -field evidence from the Okinawa Islands, Japan. Mar Geol 283:66–78. doi:10.1016/j.margeo.2010.09.007

Hall AM (2011) Storm wave currents, boulder movement and shore platform develop-
    ment: a case study from East Lothian, Scotland. Mar Geol 283:98–105. doi:10.1016/
    j.margeo.2010.10.024
Hall AM, Hansom JD, Williams DM, Jarvis J (2006) Distribution, geomorphology and lithofa-
    cies of cliff-top storm deposits: examples from the high-energy coasts of Scotland and
    Ireland. Mar Geol 232(3–4):131–155. doi:10.1016/j.margeo.2006.06.008
Hall AM, Hansom JD, Jarvis J (2008) Patterns and rates of erosion produced by high energy
    wave processes on hard rock headlands: the Grind of the Navir, Shetland, Scotland. Mar
    Geol 248(1–2):28–46. doi:10.1016/j.margeo.2007.10.007
Hall AM, Hansom JD, Williams DM (2010) Wave-emplaced coarse debris and megaclasts in
    Ireland and Scotland: boulder transport in a high-energy littoral environment: a discussion. J
    Geol 118:699–704. doi:10.1086/656357
Hansom J, Hall A (2009) Magnitude and frequency of extra-tropical North Atlantic cyclones:
    a chronology from cliff-top storm deposits. Quatern Int 195:42–52. doi:10.1016/j.quaint.
    2007.11.010
Hayne M, Chappell J (2001) Cyclone frequency during the last 5000 years at Curacoa Island,
    north Queensland, Australia. Palaeogeogr Palaeoclimatol Palaeoecol 168(3–4):207–219.
    doi:10.1016/S0031-0182(00)00217-0
Hearn CJ (2011) Hydrodynamics of coral reef systems. In: Hopley D (ed) Encyclopedia of mod-
    ern coral reefs: structure, form and process. Springer Netherlands, Dordrecht, pp 563–573
Hearty PJ (1997) Boulder deposits from large waves during the last interglaciation on North
    Eleuthera Island, Bahamas. Quat Res 48(3):326–338. doi:10.1006/qres.1997.1926
Hernandez-Avila ML, Roberts HH, Rouse LJ (1977) Hurricane-generated waves and coastal
    boulder rampart formation. In: Proceedings of the Third International Coral Reef
    Symposium, vol. 2, pp. 71–78. Miami
Imamura F, Goto K, Ohkubo S (2008) A numerical model for the transport of a boulder by tsu-
    nami. Journal of Geophysical Research 113(C1):1-12. doi:10.1029/2007JC004170
Jones B, Hunter IG (1992) Very large boulders on the coast of Grand Cayman: the effects of
    giant waves on rocky coastlines. J Coastal Res 8(4):763–774
Kelletat D, Scheffers SR, Scheffers A (2007) Field signatures of the SE-Asian mega-tsunami
    along the west coast of Thailand compared to Holocene paleo-tsunami from the Atlantic
    region. Pure Appl Geophys 164(2–3):413–431. doi:10.1007/s00024-006-0171-6
Kennedy DM, Tannock KL, Crozier MJ, Rieser U (2007) Boulders of MIS 5 age depos-
    ited by a tsunami on the coast of Otago, New Zealand. Sed Geol 200:222–231.
    doi:10.1016/j.sedgeo.2007.01.005
Kinahan GH, Leonard H, Cruise RJ (1871) Memoirs of the Geological Survey of Ireland. Sheets
    104 and 113, Dublin
Lavigne F, Paris R, Wassmer P, Gomez C, Brunstein D, Grancher D, Vautier F, Sartohadi J,
    Setiawan A, Gunawan Syahnan T, Waluyo Fachrizal B, Mardiatno D, Widagdo A, Cahyadi
    R, Lespinasse N, Mahieu L (2006) Learning from a major disaster (Banda Aceh, December
    26th, 2004): a methodology to calibrate simulation codes for tsunami inundation models.
    Zeitschrift für Geomorphologie. Supplementband 146:253–265
Liu K, Shen C, Louie K (2001) A 1,000-year history of typhoon landfalls in Guangdong,
    Southern China, reconstructed from Chinese historical documentary records. Ann Assoc Am
    Geogr 91:453–464. doi:10.1111/0004-5608.00253
Mader CL (2004) Numerical modeling of water waves, 2nd edn. CRC Press, Boca Raton
Maouche S, Morhange C, Meghraoui M (2009) Large boulder accumulation on the Algerian
    coast evidence tsunami events in the western Mediterranean. Mar Geol 262(1–4):96–104.
    doi:10.1016/j.margeo.2009.03.013
Mastronuzzi G, Sansò P (2004) Large boulder accumulations by extreme waves along the Adriatic
    coast of southern Apulia (Italy). Quatern Int 120(1):173–184. doi:10.1016/j.quaint.2004.01.016
Mastronuzzi G, Pignatelli C, Sansò P, Selleri G (2007) Boulder accumulations produced by the
    20th of February, 1743 tsunami along the coast of southeastern Salento (Apulia region,
    Italy). Mar Geol 242(1–3):191–205. doi:10.1016/j.margeo.2006.10.025

McAdoo BG, Ah-Leong JS, Bell L, Ifopo P, Ward J, Lovell E, Skelton P (2011) Coral reefs as buffers during the 2009 South Pacific tsunami, Upolu Island. Samoa. Earth-Sci Rev 107(1–2):147–155. doi:10.1016/j.earscirev.2010.11.005

Medina F, Mhammdi N, Chiguer A, Akil M, Jaaidi EB (2011) The Rabat and Larache boulder fields; new examples of high-energy deposits related to storms and tsunami waves in north-western Morocco. Nat Hazards 59:725–747. doi:10.1007/s11069-011-9792-x

Mhammdi N, Medina F, Kelletat D (2008) Large boulders along the Rabat coast (Morocco); possible emplacement by the November, 1st, 1755 AD tsunami. Sci Tsunami Hazards 27:17–30

Morton RA, Richmond BM, Jaffe BE, Gelfenbaum G (2006) Reconnaissance investigation of Caribbean extreme wave deposits- preliminary investigations, interpretations, and research directions. Open-file report 2006-1293, USGS

Nandasena NAK, Paris R, Tanaka N (2011a) Numerical assessment of boulder transport by the 2004 Indian Ocean tsunami in Lhoknga, West Banda Aceh (Sumatra, Indonesia). Comput Geosci 37:1391–1399. doi:10.1016/j.cageo.2011.02.001

Nandasena NAK, Paris R, Tanaka N (2011b) Reassessment of hydrodynamic equations: Minimum flow velocity to initiate boulder transport by high energy events (storms, tsunamis). Mar Geol 281:71–84. doi:10.1016/j.margeo.2011.02.005

Nichols G (1999) Sedimentology and Stratigraphy, 1st edn. Wiley-Blackwell, West Sussex

Noji M, Imamura F, Shuto N (1985) Numerical simulation of movement of large rocks transported by tsunamis. In: Proceedings of the IUGG/IOC International Tsunami Symposium, pp. 189–197

Noormets R, Felton EA, Crook KAW (2002) Sedimentology of rocky shorelines: 2. Shoreline megaclasts on the north shore of Oahu, Hawaii- origins and history. Sed Geol 150(1–2): 31–45. doi:10.1016/S0037-0738(01)00266-4

Noormets R, Crook KAW, Felton EA (2004) Sedimentology of rocky shorelines: 3. Hydrodynamics of megaclast emplacement and transport on a shore platform, Oahu. Hawaii Sediment Geol 172:41–65. doi:10.1016/j.sedgeo.2004.07.006

Nott J (1997) Extremely high-energy wave deposits inside the Great Barrier Reef, Australia: determining the cause- tsunami or tropical cyclone. Mar Geol 41(1–4):193–207. doi:10.1016/S0025-3227(97)00063-7

Nott J (2000) Records of prehistorical tsunamis from boulder deposits, evidence from Australia. Sci Tsunami Hazards 18(1):3–14

Nott J (2003) Waves, coastal boulder deposits and the importance of the pre-transport setting. Earth Planet Sci Lett 210(1–2):269–276. doi:10.1016/S0012-821X(03)00104-3

Nott J (2004) The tsunami hypothesis: comparisons of the field evidence against the effects, on the Western Australian coast, of some of the most powerful storms on Earth. Mar Geol 208:1–12. doi:10.1016/j.margeo.2004.04.023

Nott J, Bryant E (2003) Extreme marine inundations (tsunamis?) of coastal Western Australia. J Geol 111(6):691–706. doi:10.1086/378485

Omoto K (2010) Characteristics of travertine terraces, tsunami boulders and cemented beach sand beds observed on southeast coast of Miyako Island, southwest of Japan. Proceedings of the Institute of Natural Sciences, Nihon University, No. 45

Paris R, Wassmer P, Sartohadi J, Lavigne F, Barthomeuf B, Desgages E, Grancher D, Baumert P, Vautier F, Brunstein D, Gomez C (2009) Tsunamis as geomorphic crises: Lessons from the December 26, 2004 tsunami in Lhok Nga, West Banda Aceh (Sumatra, Indonesia). Geomorphology 104:59–72. doi:10.1016/j.geomorph.2008.05.040

Paris R, Fournier J, Poizot E, Etienne S, Morin J, Lavigne F, Wassmer P (2010) Boulder and fine sediment transport and deposition by the 2004 tsunami in Lhok Nga (western Banda Aceh, Sumatra, Indonesia): a coupled offshore- onshore model. Mar Geol 268:43–54. doi:10.1016/j.geomorph.2009.02.008

Paris R, Naylor LA, Stephenson WJ (2011) Boulders as a signature of storms on rock coasts. Mar Geol 283:1–11. doi:10.1016/j.margeo.2011.03.016

Pignatelli C, Sansò P, Mastronuzzi G (2009) Evaluation of tsunami flooding using geomorphologic evidence. Mar Geol 260(1–4):6–18. doi:10.1016/j.margeo.2009.01.002

Rahiman TIH, Pettinga JR, Watts P (2007) The source mechanism and numerical modelling of the 1953 Suva tsunami, Fiji. Mar Geol 237(1–2):55–70. doi:10.1016/j.margeo.2006.10.036

Regnauld H, Oszwald J, Planchon O, Pignatelli C, Piscitelli A, Mastronuzzi G, Audevard A (2010) Polygenetic (tsunami and storm) deposits? A case study from Ushant Island, western France. Zeitschrift für Geomorphologie N.F. 54(Supplementband 3):197–217. doi:10.1127/0372-8854/2010/0054S3-0000

Richmond BM, Watt S, Buckley M, Jaffe BE, Gelfenbaum G, Morton RA (2011a) Recent storm and tsunami coarse-clast deposit characteristics, southeast Hawaii. Mar Geol 283(1–4): 79–89. doi:10.1016/j.margeo.2010.08.001

Richmond BM, Buckley M, Etienne S, Chagué-Goff C, Clark K, Goff J, Dominey-Howes D, Strotz L (2011b) Deposits, flow characteristics, and landscape change resulting from the September 2009 South Pacific tsunami in the Samoan Islands. Earth Sci Rev 107(1–2): 38–51. doi:10.1016/j.earscirev.2011.03.008

Robinson E, Rowe DC, Khan SA (2005) Mystery boulders at Galina Point. The Gleaner for October 13, 2005

Scheffers A (2002) Paleotsunami evidences from boulder deposits on Aruba, Curaçao and Bonaire. Sci Tsunami Hazards 20(1):26–37

Scheffers A (2008) Tsunami boulder deposits. In: Shiki T et al (eds) Tsunamiites—features and implications. Elsevier, Amsterdam

Scheffers A, Kelletat D (2003) Sedimentologic and geomorphologic tsunami imprints world-wide- a review. Earth Sci Rev 63(1–2):83–92. doi:10.1016/S0012-8252(03)00018-7

Scheffers A, Scheffers S, Kelletat D (2005) Paleo-tsunami relics on the southern and central Antillean Island Arc. J Coastal Res 21(2):263–273. doi:10.2112/03-0144.1

Scheffers A, Scheffers S, Kelletat D, Browne T (2009a) Wave-emplaced coarse debris and meg-aclasts in Ireland and Scotland: boulder transport in a high-energy littoral environment. J Geol 117(5):553–573. doi:10.1086/600865

Scheffers A, Scheffers S, Mastronuzzi G (2010) Assessment of extreme wave flood-ing from geomorphologic evidence in Bonaire (Netherlands Antilles). Zeitschrift für Geomorphologie N.F. 54 (Supplementband 3):219–245. doi:10.1127/0372-8854/2010/0054S3-0026

Scheffers SR, Haviser J, Browne T, Scheffers A (2009b) Tsunamis, hurricanes, the demise of coral reefs and shifts in prehistoric human populations in the Caribbean. Quatern Int 195 (1–2):69–87. doi:10.1016/j.quaint.2008.07.016

Schellmann G, Beerten K, Radtke U (2008) Electron spin resonance (ESR) dating of Quaternary materials. Eiszeitalter und Gegenwart Quatern Sci 57(1–2):150–178

Scicchitano G, Monaco C, Tortorici L (2007) Large boulder deposits by tsunami waves along the Ionian coast of south-eastern Sicily (Italy). Mar Geol 238(1–4):75–91. doi:10.1016/j.margeo.2006.12.005

Shah-hosseini M, Morhange C, Beni AN, Marriner N, Lahijani H, Hamzeh M, Sabatier F (2011) Coastal boulders as evidence for high-energy waves on the Iranian coast of Makran. Mar Geol 290(1–4):17–28. doi:10.1016/j.margeo.2011.10.003

Sharp BB, Sharp DB (1996) Water hammer: practical solutions. Elsevier Science, Oxford, p 192

Simkin T, Fiske RS (1983) Krakatau, 1883: the volcanic eruption and its effects. Smithsonian Institute Press, Washington, D.C

Spiske M, Bahlburg H (2011) A quasi-experimental setting of coarse clast transport by the 2010 Chile tsunami (Bucalemu, Central Chile). Mar Geol 289(1–4):72–85. doi:10.1016/j.margeo.2011.09.007

Spiske M, Böröcz Z, Bahlburg H (2008) The role of porosity in discriminating between tsunami and hurricane emplacement of boulders: a case study from the Lesser Antilles, southern Caribbean. Earth Planet Sci Lett 268:284–396. doi:10.1016/j.epsl.2008.01.030

Stoddart DR (1969) Reconnaissance geomorphology of Rangiroa Atoll, Tuamotu Archipelago. Atoll Res Bull 125:1–32

Sussmilch CA (1912) Note on some recent marine erosion at Bondi. J Proc R Soc NSW 46:155–158

Suzuki A, Yokoyama Y, Kan H, Minoshima K, Matsuzaki H, Hamanaka N, Kawahata H (2008) Identification of 1771 Meiwa Tsunami deposits using a combination of radiocarbon dating and oxygen isotope microprofiling of emerged massive Porites boulders. Quat Geochronol 3(3):226–234. doi:10.1016/j.quageo.2007.12.002

Switzer AD, Burston JM (2010) Competing mechanisms for boulder deposition on the southeast Australian coast. Geomorphology 114(1–2):42–54. doi:10.1016/j.geomorph.2009.02.009

Taggart BE, Lundberg JI, Carew JL, Mylroie JE (1993) Holocene reef-rock boulders on Isla de Mona, Puerto Rico: transported by a hurricane or seismic sea wave. Geological Society of America Annual Meeting, Abstracts with Program 25:61

Talling PJ, Wynn RB, Masson DG, Frenz M, Cronin BT, Schiebel R, Akhmetzhanov AM, Dallmeier-Tiessen S, Benetti S, Weaver PPE, Georgiopoulou A, Zühlsdorff C, Amy LA (2007) Onset of submarine debris flow deposition far from original giant landslide. Nature 450(7169):541–544. doi:10.1038/nature06313

Tappin DR, Evans HM, Jordan CJ, Richmond B, Sugawara D, Goto K (2012) Coastal changes in the Sendai area from the impact of the 2011 Tōhoku-oki tsunami: Interpretations of time series satellite images, helicopter-borne video footage and field observations. Sed Geol 282:151–174. doi:10.1016/j.sedgeo.2012.09.011

Terry JP (2007) Tropical cyclones: climatology and impacts in the South Pacific. Springer, New York

Terry JP, Etienne S (2010a) "Stones from the dangerous winds": reef platform mega-clasts in the tropical Pacific Islands. Nat Hazards 56(3):567–569. doi:10.1007/s11069-010-9697-0

Terry JP, Etienne S (2010b) Recent heightened tropical cyclone activity east of 180° in the South Pacific. Weather 65(7):193–195. doi:10.1002/wea.542

Umbgrove JHF (1947) Coral reefs of the East Indies. Geol Soc Am Bull 58:729–778

Walker MJC (2005) Quaternary dating methods. John Wiley and Sons Ltd, West Sussex

Watt SG, Jaffe BE, Morton RA, Richmond BM, Gelfenbaum G (2010) Description of extreme-wave deposits on the northern coast of Bonaire, Netherlands Antilles. USGS Open-file report 2010-1180

Weiss R (2012) The mystery of boulders moved by tsunamis and storms. Mar Geol 295–298: 28–33. doi:10.1016/j.margeo.2011.12.001

Williams D, Hall AM (2004) Cliff-top megaclast deposits of Ireland, a record of extreme waves in the North Atlantic—storms or tsunamis? Mar Geol 206(1–4):101–117. doi:10.1016/j.margeo.2004.02.002

Young RW, Bryant EA, Price DM (1996) Catastrophic wave (tsunami?) transport of boulders in southern New South Wales, Australia. Zeitschrift für Geomorphologie 40:191–207

Yu K, Zhao J, Collerson KD, Shi Q, Chen T, Wang P, Liu T (2004) Storm cycles in the last millennium recorded in Yongshu Reef, southern South China Sea. Palaeogeogr Palaeoclimatol Palaeoecol 210(1):89–100. doi:10.1016/j.palaeo.2004.04.002

Yu K, Zhao J, Shi Q, Meng Q (2009) Reconstruction of storm/tsunami records over the last 4000 years using transported coral blocks and lagoon sediments in the southern South China Sea. Quatern Int 195(1–2):128–137. doi:10.1016/j.quaint.2008.05.004

Yu K, Zhao J, Roff G, Lybolt M, Feng Y, Clark T, Li S (2012) High-precision U-series ages of transported coral blocks on Heron Reef (southern Great Barrier Reef) and storm activity during the past century. Palaeogeogr Palaeoclimatol Palaeoecol 337–338:26–36. doi:10.1016/j.palaeo.2012.03.023

Zhao J, Neil DT, Feng Y, Yu K, Pandolfi JM (2009) High-precision U-series dating of very young cyclone-transported coral reef blocks from Heron and Wistari reefs, southern Great Barrier Reef, Australia. Quatern Int 195(1–2):122–127. doi:10.1016/j.quaint.2008.06.004

Zhao JX, Xia QK, Collerson KD (2001) Timing and duration of the last interglacial inferred from high resolution U-series chronology of stalagmite growth in southern hemisphere. Earth Planet Sci Lett 184:635–644. doi:10.1016/S0012-821X(00)00353-8

Chapter 4
Uncertainties and Continuing Challenges with Interpreting Coastal Boulders

Abstract The significance of coastal boulders for high-energy marine inundation (HEMI) studies relies on the primary control of two influential environmental parameters: mechanisms of boulder generation and the source of the boulders themselves. Uncertainties inherent to natural boulders can occasionally be overcome by examining 'anthropogenic boulders' sourced from engineered coastal defence structures. However, distinguishing the very nature of HEMI events, i.e. whether storm or tsunami in origin, is still a contentious issue. Similarly, the intrinsic ability of boulder deposits to exhaustively capture all inundation events is highly debatable: as a resilient object, a boulder can be (re)mobilised by several successive events, yet at the same time, its gradual degradation precludes it from accurately recording the oldest (original) event that emplaced it. Future progress should be achieved through an improvement and a standardisation in the collection and presentation of coastal boulder data.

4.1 Introduction

Evaluating the characteristics of (unrecorded) past HEMI events on affected coastlines is undeniably a formidable task for the coastal geomorphologist. Yet from what has been presented in the previous chapter, it is clear that information derived from coastal boulders sheds an illuminating beacon of light on such investigations. Nevertheless, in spite of the possibilities that exist, a range of obstacles, uncertainties and challenges still dog both the initial identification and later interpretation of coastal boulder deposits. These need to be recognised, addressed and then hopefully overcome through further scientific attention and research. Some of the most important issues are elaborated in the following sections.

4.2 Mechanisms of Coastal Boulder Generation

Although boulder deposition on coastlines can be caused by HEMI events, a number of other processes may also be responsible. McKenna et al. (2011) listed four additional mechanisms for coastal boulder emplacement besides HEMI events. These are:

1. Gravity-induced slope movements, dominated by rockfall from coastal cliffs (Fig. 4.1a).

Fig. 4.1 Coastal boulder origins other than marine inundation events. **a** Basaltic boulders lying on a chalky shore platform, Ballycastle, Northern Ireland. Boulders are delivered by rockfall from the Tertiary Lower Basalts lava flow (*L*) overlying the Cretaceous Upper Chalk unit (*C*) (Photo by S. Etienne 1995); **b** Uplifted coral terraces on the south coast of Erromango Island, Vanuatu (Photo by Shane Cronin, September 2003, used with permission). Quaternary coastal deposits have been uplifted at accelerating rates of 0.35–1 mm/year over the past 320,000 years (Neef and Hendy 1988); **c** Isolated large block composed of bedded carbonates near Surigao on the northern coast of Mindanao Island in The Philippines. The origin of the block is unknown, but it is probably a residual outlier remaining after erosion of a coastal outcrop. Similar features that are still connected to the rock coast occur nearby (Photo by Fernando Siringan, used with permission); **d** Coastal boulder in the Kongsfjorden area, Spitsbergen. This boulder is a glacial erratic deposited by the Conway Glacier during its post-Little Ice Age retreat. Glacial erratics are boulders transported by moving ice or floating ice rafts. Moving ice is capable of transporting much coarser clasts than flowing water. On tropical coasts, glacial processes can be ruled out as the mechanism responsible for coastal boulder deposition. (Photo by Franck Delbart, July 2002, used with permission)

2. Sea-level change, where boulders were emplaced by marine processes at an earlier time of elevated sea level and left stranded by subsequent sea-level fall (i.e. emergence). For example, during the last interglacial warm period which reached its peak 125,000 years BP, global sea level rose to approximately 6 m above its present height. Last interglacial coastlines were therefore located farther inland from their current position. Consequently, any boulders deposited on offshore submarine platforms during the interglacial marine high-stand may appear as coastal boulders today. A similar mechanism could also affect unstable landmasses, i.e. where tectonic or isostatic activity alters the relative

sea-level position (Fig. 4.1b). For example, the Huon Peninsula in Papua New Guinea has a very high rate of tectonic uplift, reaching 3.3 mm/yr during the last 120 kyr at Bobo (Chappell et al. 1996). There, marine isotope stage 5e coral terraces are today found at over 400 m a.s.l.

3. In situ exhumation of boulders at the coast from heavily-weathered bedrock by normal wave action and shoreline retreat (Figs. 4.1c and 1.1).

4. Glacial processes, where boulders were either deposited as glacial erratics (Fig. 4.1d) produced by ice shoving and ice abrasion at the shoreline, or released in situ from an eroding glacial deposit such as a moraine.

Insofar as explaining the details of these various processes falls beyond the scope of this volume, it is nonetheless important that they be recognised by coastal researchers, as this lessens the chance of coastal boulder deposits generated by these alternative mechanisms being misidentified as the result of HEMI events. On tropical coastlines, the possibility of boulder production by glacial processes can clearly be excluded. However, boulders generated or emplaced on coasts by the other mechanisms listed above may still be usefully examined, if they are later subjected to further transport by marine inundation, and especially if pre-event positions are known. This idea also applies to terrestrial boulders delivered to coasts by volcanic eruptions or fluvial processes.

4.3 Identifying Original Sources for Carbonate Boulders on Reefs

In the case of carbonate boulders deposited on reefs and beaches in tropical regimes, various sources are possible, with Paris et al. (2010) pointing out that the size and spatial distributions of coastal boulders depends heavily on the location of boulder sources. Consequently, it is imperative to distinguish between boulders that were produced by quarrying at the modern reef during HEMI events and those derived from other sources, of which several exist. Even when reef-platform coral boulders (RPCBs) are the result of wave erosion on living reefs, their exact source is not always known (Fig. 4.2), since they may have been newly eroded from the reef flat or the reef crest itself. According to Done (1992, p. 859), who favoured a reef attrition model, cyclone waves are able to "exfoliate reefs, chunk by chunk". However, boulders may also be dredged up from fore-reef slopes or offshore reef ledges if accumulated there at earlier times (Goto et al. 2010a).

Submarine work is sometimes needed for the identification of talus on reef slopes (Fig. 4.3), or for checking the seaward-facing reef edge for the existence of fractures and other damage to the normal spur-and-groove morphology, as this can provide clues as to the origin of fresh carbonate boulders. Coral ecology can then be used to identify the depth of boulder origins because scleractinian corals are highly sensitive to light availability, which diminishes with water depth. Based on coral assemblages, Goto et al. (2007) were able to infer that all coral boulders

Fig. 4.2   Reef-platform coral boulders (RPCBs) on Tetiaroa Atoll, French Polynesia, April 2010. Older boulders appear blackened due to weathering and a coating of algae, contrasting with fresh boulders produced by Tropical Cyclone Oli in February 2010. The image highlights the difficulties with determining the precise origin of new boulders if erosion scars are not clearly visible on the reef platform. (Photo by S. Etienne 2010)

Fig. 4.3   Submarine work is sometimes required to identify various reef sources of carbonate boulders, or depositional sites post HEMI event. Pictured is a massive Porites colony overturned on the fore-reef slope on Tahiti during Tropical Cyclone Oli in February 2010. (Photo by S. Etienne, April 2010)

encountered on the reef flat of Pakarang Cape, Thailand, originated from the reef edge to <10 m water depth because they were fragments of reefrock with accreted coral colonies consisting of *Porites lutea*, *Galaxea astreata*, *Favia* sp., *Coeloseris mayeri*, *Platygyra daedalea*, and *Leptoria phrygia*. These coral colonies are found at shallow depths, less than 5–10 m.

Coastal cliffs comprising raised Quaternary reef limestones may also deliver sizeable fragments onto a modern fringing reef surface by rockfall and cliff collapse. Gravity and prolonged coastal erosion over geomorphic timescales play important roles in these mass movements, as well as isolated extreme events on occasion. Niue Island (19.1°S, 169.6°W) in the central South Pacific serves as a good illustration. Niue is an isolated high carbonate island, formed over the past half a million years by the tectonic uplift of a coral-capped submarine volcano (Terry and Nunn 2003). Owing to the uplift, a sheer carbonate cliffline comprising ancient corals, now emerged, is observed around most of the island, rising some tens of metres above modern sea level (Terry 2004). Undercutting by wave action has produced deep wave-cut notches and overhangs at the base of the cliffs, which are consequently prone to failure. Through time, regular collapse delivers much coarse carbonate debris onto the narrow fringing reef. Such a scenario is common on many similar emerged carbonate islands across the Pacific. Cliff collapse and consequent carbonate boulder production may also occur episodically during specific high-energy events. On January 2004, Tropical Cyclone Heta passed within 30 km of the west coast of Niue and pounded the cliffline with violent waves. Sections of cliffline retreated by several metres through collapse, so producing many new clasts several metres in diameter that now rest on the modern narrow reef flat (Fig. 4.4). Kogure and Matsukura (2010) identified similar sources for reef-platform carbonate boulders on Kuroshima Island in Japan.

Former coral reef surfaces that grew up to meet an earlier marine high stand may be exposed later by falling eustatic sea levels. In this way, in situ breakdown of bedrock may produce carbonate boulders through solution weathering and erosion. Over time, remnants of emerged reef terraces are transformed into angular, mushroom-shaped, isolated pinnacles of dead corals, commonly known as *feo* throughout Polynesia. The coral pinnacles (*feo*) may have their tops exposed above sea level but are still attached to the modern reef at their base (see Chap. 2). If these mushroom rocks become detached from their base through wave erosion, they topple onto the modern reef surface and then take on a similar appearance to coral boulders emplaced by inundation. Clearly, this mixture of possible sources for carbonate boulders seen on reef platforms adds to the existing dilemmas of correctly interpreting the nature of past HEMI events.

A third possibility is that carbonate slabs may be derived from exhumed beachrock outcrops, where overlying beach sand has been removed. Beachrock is calcarenite rock, i.e. a type of sandstone comprising calcareous fragments, mostly shell pieces and coralline sands and gravels, although sometimes also containing pebbles of local terrestrial lithologies. Beachrock forms within the beach profile, with lithification of sediments occurring through percolating freshwater cementing together the constituent carbonate grains. If the beach

Fig. 4.4 Exposed cliffs of uplifted reef limestone along the coast of Niue Island in the central South Pacific. The erosional scar in the centre of the image reveals light-coloured unweathered dolomitic limestones, contrasting strongly with dark weathered rocks on either side. Cliff collapse was the result of buffeting by monstrous waves during Tropical Cyclone Heta in January 2004. Debris from the collapse exists as a group of fresh carbonate boulders now resting on the narrow fringing reef platform in the foreground. (Photo by Michael Bonte, 2004, used with permission)

profile above is subsequently eroded, then exposed surfaces of beachrock become pitted and scalloped by solution weathering. Outcropping beachrock then becomes available as a source of rock that may be broken up during HEMI events to produce loose slabs (Fig. 4.5); detached beachrock clasts may be pushed onshore by wave runup or dragged backwards to the intertidal zone by backwash currents.

4.4 'Anthropogenic Boulders': Advantages of Studying Boulders Sourced from Rip-Rap

If identifying the source location of coral boulders is difficult, then reconstructing palaeo-event hydrodynamics from such deposits obviously becomes a more complicated task as a consequence. Occasionally, anthropogenic structures built as sea defences, such as rock armour, may act as a coastal boulder source, thereby aiding post-event transport modelling. In Sumatra, Indonesia, Paris et al. (2009) studied a boulder field of calcareous blocks extracted from a seawall by the 2004 Indian Ocean tsunami. More than 1000 boulders were removed from the seawall and drifted landward by significant distances (up to 200 m). No fining trends were observed and their distribution suggests little rearrangement by the tsunami outflow. At Satitoa on Upolu Island in Samoa (14°02′S, 171°26′W),

Fig. 4.5 Transported beachrock slabs (flat clasts) amongst other types of boulders on the fringing reef at Dabaisha, west coast of Lu Dao Island in south east Taiwan (22°38.3′N 121°29.6′E). Some new slabs were detached from the exposed beachrock body by large waves during Typhoon Tembin in August 2012. These brown coloured slabs are easily identifiable as they stand out well against pre-existing green slabs that have a surface cover of seaweed. Numbers of beachrock fragments were pushed landwards up the beach (slope angle 10°) by wave runup, while others were pulled seawards by vigorous backwash onto the intertidal flat. One advantage of investigating the distribution of fresh beachrock clasts is that their source location may sometimes be determined with pinpoint precision from their 'jigsaw fit' into proximal in situ beachrock outcrops. The beachrock source is seen in the right of the photograph. This allows accurate measurement of transport distance, if the specific HEMI event that produced them is known (see Etienne and Terry 2012). (Photo by J. Terry, November 2012)

rip-rap boulders were carried landward during the 2009 South Pacific tsunami and were lain down inland on the coastal plain (Etienne et al. 2011a). Of importance here was that the boulder source could be precisely determined and the transport distance measured. All boulders were deposited within 162 m from the revetment (Fig. 4.6); distances were not measured shore normal, but conforming to the mean tsunami flow direction, as inferred from flow indicators in the field (Richmond et al. 2011). Likewise, concrete slabs detached from a seawall by the 2011 Tōhoku tsunami in northern Japan were examined by Goto et al. (2012) for similar purposes.

However, the use of anthropogenic boulders encounters limitations due to the restricted volume of each artificial boulder. In the Samoan example, nearly all of the boulders comprising the rock armour were carried inland by the 2009 tsunami, meaning that the transport capacity of the waves was greater than can be deduced from the measured boulders themselves. As no boulder was larger than 1 m^3,

Fig. 4.6 Rip-rap boulder field, dispersed over the coastal plain by the 2009 South Pacific tsunami at Satitoa on Upolu Island in Samoa (14°02′S, 171°26′W). *Top*: yellow spots indicate the location of the basaltic boulders; blue arrows the main tsunami direction; a–b black segment indicatesthe boulder source (rock armour). *Below*: aerial oblique picture of the area 1 day after the tsunami, New Zealand Air Force, 30 Sept 2009

the maximum size transported was limited by their original availability at the source (i.e. the rock armour). At the opposite end of the spectrum, other types of engineered coastal defences are not always useful for post-event hydrodynamic transport reconstruction. During the 2011 Tokōku tsunami for example, beach-face protection such as tetrapods were largely unaffected as most of them remained in place (Tappin et al. 2012).

4.5 Distinguishing Between Storm and Tsunami Boulders

Even in situations when available evidence suggests that coastal boulders were in fact delivered by marine inundation events rather than by other geomorphic processes, the question that naturally follows is whether storm or tsunami waves were responsible? Indeed the 'storm versus tsunami' conundrum has occupied something of a centre stage position in coastal geomorphology in recent times and remains a contentious issue. One regional illustration of this is the controversy that has surrounded the beleaguered Australian megatsunami hypothesis (AMH), which has previously fuelled intense debate in the scientific literature but is yet to be unambiguously resolved. The AMH is based on, amongst other erosional and depositional features, anomalous bouldery material found on coastal rock platforms, some occurring as stacked and imbricated debris. The rock platforms are at elevations apparently beyond the limits of modern storm waves at a number of locations along the New South Wales coastline of Australia (see Bryant et al. 1992; Bryant 2001; Bryant and Nott 2001 for proponents of the AMH, and the following work for critiques and alternative views: Felton and Crook 2003; Saintilan and Rogers 2005; Goff and Dominey-Howes 2009; Switzer and Burston 2010). A comprehensive review explaining the evolution of the vexed AMH question is provided by Courtney (2012).

Attempts to make progress forward in this arena has prompted various workers to focus efforts on establishing certain criteria for discriminating between tsunami and storm deposits (e.g. Nott 2000; Goto et al. 2010b; Scheffers et al. 2009; Lorang 2011). While Nott (1997, 2000) considered that separating storm from tsunami boulders can be accomplished using hydrodynamic transport equations, other researchers believe that distinguishing between these two mechanisms is not so straightforward (Switzer and Burston 2010). It is commonly thought that large tsunamis possess much greater wave energy than the strongest possible storms, mainly due to their longer wave period and duration (Nott 2003; Scheffers et al. 2009). Accepting this assumption, it follows that tsunami waves should therefore be more effective in detaching and transporting boulder-sized material (Noormets et al. 2004).

By appreciating observable distinctions in wave periodicity, where wind-generated wave periods lie between 10–40 s while tsunami wave periods span tens of minutes to hours, Lorang (2011) developed an equation (Box 5) to distinguish storm from tsunami boulders by calculating the period of the waves responsible for their deposition. Investigating boulders at Kalalau beach in Hawaii as a case study, the author demonstrated how this 'wave-competence' approach may be applied. However, to determine values for the equation parameters, Lorang (2011) used the earlier numerical methods for estimating maximum swash velocity of Nott (2003), in spite of existing rebuttals against the embedded concept that tsunami waves are four times more capable of boulder transportation than storms (Morton et al. 2006).

Box 5
Equation to Distinguish Storm from Tsunami Boulders
(Lorang 2011)

$$T = 2 \bigg/ g \left(\frac{\rho}{\rho_s - \rho}\right) \left(\frac{C_d}{S}\right) \left(\frac{h_{clast}}{D_i}\right) U_{max}$$

Parameters:

T	wave period (s)
G	gravitational acceleration (= 9.81 m/s^2)
ρ_s	density of boulder (kg/m^3)
ρ	water density (kg/m^3)
S	slope of the beach face (angle in radians)
h$_{clast}$	height of clast above original elevation (m)
D$_i$	intermediate diameter of boulder (m)
U$_{max}$	maximum swash or bore velocity (m/s)

Refer to Lorang (2011) for the steps, equations and values for deriving U$_{max}$.

Alternatively, a number of publications have suggested that a so-called 'transport figure' (refer to Sect. 3.3.1) can be used as a proxy measure of wave energy in order to differentiate between tsunami and storm boulders (Scheffers and Kelletat 2003; Scheffers and Scheffers 2007; Etienne et al. 2011a). Apparently, transport figures can attain values up to 11×10^6 for tsunami-generated deposits, while they seldom exceed 5×10^3 for storms (Scheffers and Kelletat 2003; Etienne and Paris 2010). Yet, in calculations the horizontal transport distance between the source area and the depositional site of a boulder needs to be known. This means that estimating transport distance becomes problematic if the boulder source is uncertain (Fig. 4.7), or if more than one phase of boulder remobilisation has occurred (see Sect. 4.7).

Generally speaking, tsunamis are often favoured over storms as responsible for boulder production. The premise for this is a simple one based on the abundance of historical information for the latter: it is believed that the temporal extent of storm records now provides a long enough dataset for establishing the near-maximum possible intensity of storms (Frohlich et al. 2011). Thus, at individual sites where the hypothetical maximum storm intensity proves insufficient to generate waves capable of throwing up the largest observed coastal boulders, tsunamis must necessarily be evoked by default as the only possible mechanism for boulder emplacement.

It is important to recognise that single boulders do not provide sufficient information for identifying the true nature of a HEMI event. Sedimentary (boulder) assemblages, if available, provide much more convincing evidence for deciphering storm and tsunami characteristics. For example, Etienne and Paris (2010) advocate that constructional landforms such as boulder ridges with imbricated clasts are of storm origin, since they reveal the accretionary work of multiple successive waves or events. In contrast, no boulder ridges have yet been observed after the passage of a modern tsunami on a low-lying coastal plain, although concentrations

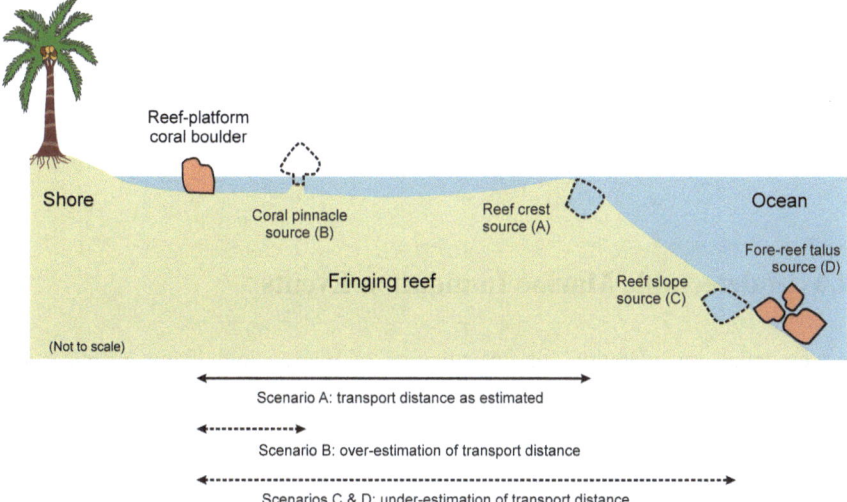

Fig. 4.7 Potential problems with determining the transport distance of a new reef-platform coral boulder (RPCB) during the HEMI event that produced it, associated with identifying the original source location. The reef crest is the most common source for RPCBs, therefore the distance between the boulder location and the reef crest is normally assumed to be the transport distance (*scenario A*). However, if the RCPB was produced instead by the erosion of a coral pinnacle (*feo* or remnant feature of older emerged reef), the actual transport distance is less than the measured distance (*scenario B*). This would lead to an over-estimation of HEMI event energy. On the contrary, if the boulder was extricated from the fore-reef slope or dredged up from pre-existing talus on the reef slope, the wave energy will be under-estimated because the actual transport distance is longer than assumed (*scenarios C* and *D*). Moreover, higher energy is required to extract and transport a boulder from a submarine source (reef slope or talus accumulation) because vertical lifting is required, compared to reef flat and reef crest sources that are at or near the sea surface

of boulders can form along the shoreline where topographic obstacles such as sand dunes drastically reduce the current velocity (Goto et al. 2010c). Dispersed boulder fields have been observed after the 2004 Indian Ocean tsunami (Paris et al. 2007, 2009), the 2009 South Pacific tsunami (Etienne et al. 2011a) and the 2011 Tōhoku tsunami (Goto et al. 2012). Goto et al. (2010c) saw a landward fining of boulders related to storm-wave activity, whereas such a trend is notably absent in historical tsunami boulder deposits in the Ryukyu Islands of southern Japan. However, recent studies on the 2011 Tōhoku tsunami show opposing results, with both landward coarsening and fining trends at Sabusawa Island in northern Japan (Goto et al. 2012). The important difference in this case is that the Sabusawa Island study area is surrounded by low hills, thereby restricting inundation to a narrow coastal plain. The drainage route for return flow was therefore confined to an erosional channel through the broken seawall. As a result, a far less complex pattern of inundation occurred than might otherwise have been expected across a wide, flat lowland.

Bearing in mind the points above, it appears wise to remain cautious. An unequivocal, universal signature in boulder deposits that allows discrimination between storm and tsunami mechanisms for their emplacement has yet to be conceived and is likely to remain elusive for some time to come. As outlined, this is partly because the characteristics of boulder deposits are both site- and source-specific. This conclusion should not altogether come as a surprise, as the situation is similar with fine-grained sediments laid down by marine inundation events (Kortekaas and Dawson 2007; Engel et al. 2010).

4.6 Undetectable Marine Inundation Events

An essential consideration for investigators of marine inundation events is that not all storms or tsunamis leave behind large coastal boulders as evidence of their occurrence. For instance, Kelletat et al. (2007) noted that while the island chain west of Sumatra was affected by the magnitude M_w 8.6 Nias earthquake on 28 March 2005, the resulting tsunami was fairly small with a maximum runup of 4 m. Neither major reef destruction nor bouldery deposits were observed on Simeulue Island, even though part of this island's coastline lay within 100 km of the earthquake epicentre. Another event, the 2004 Indian Ocean tsunami (IOT), was one of the most catastrophic tsunamis in history. Close to the earthquake epicentre along the coastline of western Sumatra, Paris et al. (2009, 2010) mapped hundreds of coral boulders up to 85 tons in weight, transported both onshore and offshore by the waves. In contrast, the largest boulder in western Thailand from the IOT was discovered on the tombolo adjoining Phi Phi Island and weighed a modest 40 tons. There, the relatively low production of large clastic deposits from this otherwise disastrous tsunami prompted Kelletat et al. (2007) to suggest how other factors must significantly affect the deposition of coastal sediments. These factors included the mechanics of the tsunami-generating earthquake (slow shock impulse on water masses), earthquake-zone water depth (shallow) and bathymetry of the coast (shallow water).

The availability or condition of the boulder source is another influence, especially on the production of RPCBs. Hayne and Chappell (2001) studied tropical storm frequency from coral rubble ridges at Curacoa Island in north Queensland, Australia. They proposed that if the reef (source material) regeneration time exceeds the recurrence interval between storms, then ridges of coral detritus cannot be formed even when storm wave energy is high. Some storms are therefore undetectable if relying on sedimentary evidence alone, leading to an incomplete event history. Likewise, on tropical coastlines where fringing reefs are absent or poorly developed, the delivery of coral boulders during HEMI events will be limited.

Taken together, these considerations underscore the chief drawback with employing boulders alone to interpret HEMI-event history: boulders only provide a minimum estimation of event frequency (Yu et al. 2004, 2009). Of crucial importance also, it is clear that an absence of bouldery deposits on a particular coastline does not necessarily signify zero exposure, nor should be thought as representing a minimal risk of future HEMI events at that coastal location.

4.7 Boulder Reworking by Backwash or Subsequent Events

The idea has been advanced that size distributions of coastal boulders can help with the analysis of marine inundation processes (Goto et al. 2007). Although there are merits to this approach, one problem is that backwash or return flow sometimes displaces a boulder back towards its original location (e.g. the reef edge), thus lessening our ability to make meaningful interpretation. Numerical modelling has indicated that tsunami backwash can rearrange smaller boulders in both onshore and offshore directions (Loevenbruck et al. 2007). Field observations support this finding. In north west Morocco, for example, Medina et al. (2011) reported that coastal boulders were imbricated by tsunami backwash, as revealed by their landward-dipping orientations.

Besides backwash movement during the original HEMI event, abundant evidence also proves how wave-deposited boulders can be redistributed by subsequent inundations (Noormets et al. 2002; Stephenson and Naylor 2011; Fichaut and Suanez 2011). An example from Hawaii serves to illustrate. On Oahu, aerial photos reveal that a 96-ton limestone boulder was delivered onto a coastal rock platform between 1940 and 1950. Subsequently, the clast was twice shifted landwards over the periods 1952–1962 and 1969–1971, for a total distance of about 30 m (Noormets et al. 2002). If it were not for the availability of the aerial photos showing boulder movement history, the assumption of a single phase of transport would be an easy mistake to make, and any estimations of inundation processes or wave energy would be erroneous in consequence.

Eye-witness accounts and photographs can occasionally offer details on the movements of pre-existing coastal boulders after subsequent HEMI events, although unfortunately such information is rarely available. In response, one research group has made a call for a 'GEOBOULDER' web portal to be established (Etienne et al. 2011b; Terry et al. 2011). The proposition is for the launch of an online research tool with the key objective of centralising existing data on coastal boulders. Aiming eventually to provide global coverage, the intention is to encourage active researchers to upload available data on boulder positions, dimensions, morphometry and a range of associated environmental parameters. Future investigators of boulder reworking would then have unrestricted access to these records of pre-transport settings. Using this precise information, modelling would then be one way to improve current understanding of sediment transport processes during modern events, as well as permitting better reconstruction of palaeo-events.

Finally, while noting that an individual boulder might lead a somewhat 'turbulent' post-depositional existence, possibly experiencing multiple episodes of remobilisation, boulder assemblages on the contrary may exhibit a greater degree of permanence in the coastal landscape. One reason for this is that when boulders occur in sufficient numbers in the intertidal zone, secondary processes of fretting and fitting may 'imprison' boulders within the overall deposit (Hills 1970). Adjacent clasts may be packed into a three-dimensional interlocking mass, affording considerable stability to the fitted fabric. This can enable an entire boulder mass to resist movement even in extreme storms on exposed high-energy coastlines (Bishop and Hugues 1989), so limiting possibilities for further reworking.

4.8 Longevity of Boulders

Over time, coastal boulders suffer gradual degradation by various erosion processes typical of marine and (partially) subaerial environments. This causes their size to diminish, until the reduced rubble eventually becomes incorporated into finer sediment fractions: cobbles, pebbles and sand. Some boulders may also be broken down more rapidly by storm activity. Consequently, because old boulders are smaller than when first emplaced on coasts (the amount of decrease being time-dependent), this means that measurements of their size and volume at the present time yield lower than original values (Fig. 4.8). If these values are then substituted into calculations of wave height or transport figures, the energy characteristics of ancient HEMI events will be underestimated.

Several physical mechanisms are responsible for boulder size reduction, including inter-clast collision, crushing and breakdown in high-energy conditions (Chen et al. 2011; Stephenson and Naylor 2011). On Rangiroa Atoll in French Polynesia, for example, at least one pre-existing 'cyclopean' coral boulder on Motu Maereherehonae was broken apart by a series of cyclones in 1983 (Ricard 1985). On reef platforms generally, Scoffin (1993, p. 213) stated that "By this means the cycle of reef top sedimentation proceeds: corals grow in shallow water on the reef front, storms break the coral, transport the fragments and drop them as a ridge on the reef flat, the coarse particles are broken down to cobbles, gravel, sand and mud sizes and transported into the lagoon or off the reef. The storm ridges are a half-way house for much of the reef calcium carbonate".

Bioerosion also affects the durability of carbonate clasts, since boring plants and animals are known to transform reef materials over time into coral mud in the

Fig. 4.8 RPCB size reduction after deposition on Makemo Atoll, French Polynesia. **a** Inter-clast collision during a high-energy hydraulic event (swell, surge) is an important process of boulder recycling, but preliminary weakening of the coral fabric by bioerosion and weathering favours the process. **b** A boulder balanced on an uneven reef surface is unstable and subject to gravity tension that tends to split it into several pieces. Photos by S. Etienne, December 2011

lagoon (Hobbs 1933). Boring organisms like sponges, bivalve molluscs, marine worms, and grazers like sea urchins and crown-of-thorns starfish are common on intertidal reef flats (Scoffin 1993; Kázmér and Taboroši 2012). Observing hurricane deposits on the Low Isles of the Great Barrier Reef, Scoffin (1993) recognised a distinct zonation of bioerosion on intertidal boulders. The upper zone of RPCBs was characterised by oyster encrustations, while chitons and boring clams dominated the middle and basal zones respectively. In the supratidal area, coral debris was commonly blackened by a thin coat of endolithic filamentous algae. Broad zonation can also be observed in terms of the relative importance of solution weathering, bioerosion and physical marine abrasion, as seen in Fig. 4.9.

Coral boulders exposed above sea level during some of all of the tidal cycle are subject to solution weathering by rainwater, a weak carbonic acid. Seawater, though, is not acidic with pH often around 8.2, so bioerosion and mechanical processes (wave action) prevail in intertidal reef areas (Trudgill 2011). From investigations on Aldabra Atoll in the Indian Ocean, Trudgill (1976, 2011) showed that erosion rates on intertidal reef limestones ranged from 1 mm/year on sheltered

Fig. 4.9 One of many reef-platform carbonate boulders sitting on the wide fringing reef near the village of Bouma in eastern Taveuni Island, Fiji. Quasi-horizontal zones may be defined in terms of the relative dominance of various degradational processes operating on the boulder, all of which act to decrease its total mass and size over time. On the top part that is exposed above sea level for most of the tidal cycle (*zone A*), solution weathering by rainwater has caused minor karstification with shallow pits separated by sharp angular ridges of limestone. Barnacle encrustations are the main feature of the middle section (*zone B*). Bioerosion and marine abrasion dominate near the base (*zone C*); the lower portion of the boulder is riddled with boring worms and has a coating of algae grazed by chitons, sea urchins and various molluscs, mirroring the usual ecological activity on the surrounding reef platform. (Photo by J. Terry, June 2012)

coasts where bioerosion dominated, to 4–5 mm/year on coasts exposed to wave action. It is not unreasonable to expect comparable erosion rates to be operating on coral boulders occupying similar reef-platform environments elsewhere.

Besides natural processes of degradation, anthropogenic disturbance also plays a role on occasion in boulder longevity. Larger coastal boulders have been known to be deliberately removed because of a perceived nuisance to human activity. At Discovery Bay in Jamaica, many boulders (up to 50 tons) sit on a Pleistocene-age terrace 50 m inland from the shoreline. These were likely emplaced by ancient HEMI events at times of higher sea level. Rowe et al. (2009) noted that several were taken away to give space for coastal developments. Equally, on Tupai Atoll in French Polynesia, an 18 m^3 RPCB named *Paeotini* was dynamited at the request of aircraft pilots sometime over the past decade owing to its close proximity to the island's airstrip (Terry and Etienne 2010). Elsewhere, boulders have been collected to be used as rip-rap in shoreline defences or for other purposes. In Nha Trang in central Vietnam, for example, marine scientists at the Institute of Oceanography are aware of coral boulders on the fringing reef being 'harvested' by local cement factories (Dr. Vu Tuan Anh, personal communication to J. Terry, 2012).

4.9 Data Collection and Presentation

4.9.1 Volumetric Calculations

Frequently in coastal studies, boulder positions are first mapped using GPS devices, immediately followed by dimensional measurements. Three boulder axes are measured with the a- b- and c-axis representing the long, intermediate and short axis respectively (Fig. 4.10). Boulder volume estimation is then possible from the familiar equations for ellipsoidal or prismatic shapes (Paris et al. 2011), assuming that boulders approximately conform to these regular geometric shapes. Unfortunately, this method becomes more challenging when boulders are irregularly shaped (e.g. trapezoidal), because the actual volume deviates appreciably from the calculated volume (Fig. 4.11). It has therefore been mentioned that geometric calculations usually over-estimate the volume of boulders (Spiske et al. 2008; Watt et al. 2010).

In an attempt to rectify such errors, Engel and May (2012) employ a new technique which uses differential global positioning systems (DGPS) to measure boulder volume more realistically. For the same purpose, other researchers are currently exploring the use of three-dimensional modelling of boulder shape and volume using close-range digital photogrammetry (Fig. 4.12). This alternative methodology, based on the algorithms named Structure from Motion (SFM), is still in a developmental stage, but relies on creating precise, measurable and textured 3D models of coastal boulders without physical contact with the object. Simply stated, a digital camera is used in the field to rapidly acquire dozens of images of the object of interest. Multiple pictures (at least 30 photos) are taken in a single 'orbit' around the boulder in question, aiming towards its centre. Instant

Fig. 4.10 Example of a RPCB with an irregular shape on Makemo Atoll in French Polynesia. Only a rough estimation of this boulder's volume can be acquired by assuming it has a triangular cross-section with a rectangular base. Note that the a-, b- and c-axis refer to the long, intermediate and short axis respectively. While the height of the boulder is usually recorded as the c-axis (as shown here), there are exceptions where the width is shorter than the height. In such cases, the boulder is considered to be resting on its side rather than its base, and so the width would be marked as the c-axis. (Photo by A. Lau, December 2011)

quality control of the images is possible. Several ground control points (at least two), identified in the object space, are used to set up a baseline which will be further used to set the scale of the model. In the laboratory at a later stage, a 3D model is built up from the multi-view digital photographs. This allows reconstruction of quantifiable and realistic-looking textured images of complex-shaped boulders. Viewing is possible in 2.5D on a regular computer monitor, or in full 3D using stereoscopic monitors and glasses. The resulting models can be exported into any CAD software, measured in any dimension and include options for the calculation of surface area, volume, volumetric slices, resections and other associated variables. While building the fine model is computationally extensive, the initial model can be built in situ (i.e. in the field using a portable laptop computer) to ensure that the photographic coverage is adequate for precise model reconstruction using computationally powerful office workstations.

4.9.2 Inconsistent Data

Inconsistency with the type of coastal boulder data collected and presented by researchers is a problem encountered in the published literature. Concerning the

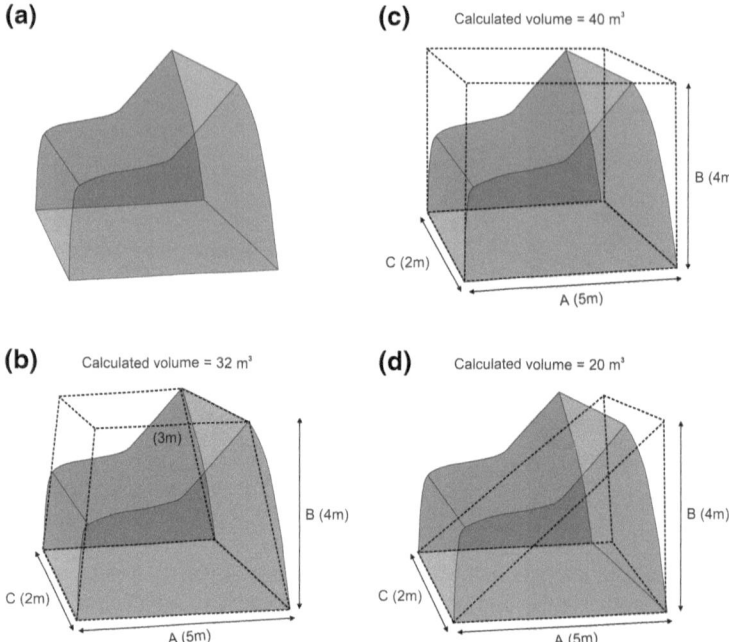

Fig. 4.11 A fictitious boulder with an irregular shape, to demonstrate several possible scenarios for volume calculation. **a** Actual shape of boulder, **b** Rectangular assumption, **c** Trapezium assumption, **d** Triangular assumption. While the rectangular and trapezium calculations yield over-estimations, the triangular assumption under-estimates the true boulder volume. In the given case, the trapezium estimation appears to provide the best choice among the three alternatives for rapid assessment of boulder volume

most basic of measurements, i.e. boulder dimensions, some studies report the a-axis length, while others use b-axis length instead. Comparison between various datasets becomes difficult as a result. Elsewhere, boulder weight is the preferred metric whereas clast size is not always mentioned (refer to latter columns of Tables 2.3 and 3.1 for examples). This situation has probably arisen due to a lack of measurement guidelines in the past. Partly in response, the review by Paris et al. (2011) advised that researchers carrying out future work on coastal boulders should stick to an agreed set of standards, in order to build more consistent and comparable datasets. Our recommendation here goes a step further, advocating that future investigations gather data on a range of mandatory parameters that are recorded in a normalised format. Quantitative, semi-quantitative and qualitative information can be incorporated (Box 6). Details on procedures have already been proposed by Etienne (2010) (Table 4.1) and are intended for inclusion in the planned GEOBOULDER web portal outlined earlier. Standardised datasets will be especially valuable as a logistical tool when used in systematic and repeated coastal surveys.

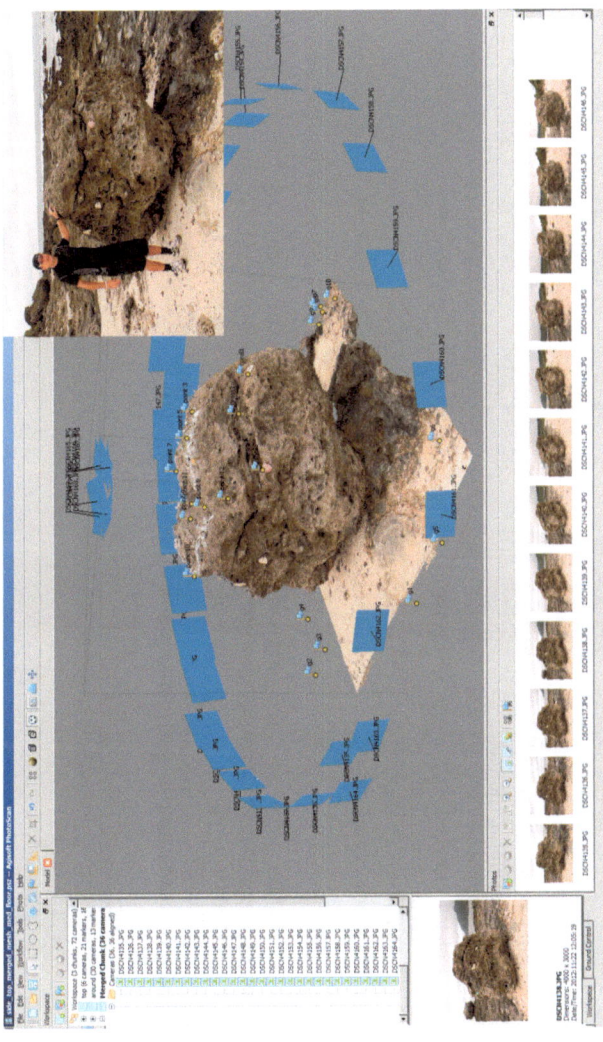

Fig. 4.12 *Inset, top right:* a carbonate boulder at a field investigation site on the south west coast of Lu Dao Island in southern Taiwan (photo by J. Terry, June 2012). The position is 22°38.190′N 121°29.673′E ± 5 m. Using a tape for crude measurements, the boulder has the following axis dimensions: a = 248 cm, b = 219 cm, c = 204 cm. This estimates boulder volume to be 11.08 m³ (rectangular assumption) or 5.86 m³ (spheroidal assumption). *Main image:* screenshot of preliminary three-dimensional model of the same boulder, using close-range photogrammetric methods. The textured model was produced from over 35 digital photographs taken in a single revolution around the object and from above. The blue rectangles show the camera position for individual photos used to build the 3D model. Calculated model volume is 4.54 m³. If a sufficient number of images are taken of the object and ground control points are recorded, then digital photogrammetry is an effective tool that enables calculation of the boulder surface area and volume more accurately than traditional measurement techniques. Image courtesy of Dr Gennady Gienko, University of Alaska, Anchorage, unpublished data (Gienko and Terry, forthcoming)

Table 4.1 Summary environmental descriptors of coastal boulders, recommended as mandatory parameters to be recorded in the field in a standard and normalised dataset [expanded from original version by Etienne(2010)]

Descriptor	Observation or measurement	Method or tool	Information type[1]	Scale or precision	Interpretation or value
Geographical location[2]					
Location	Absolute position	GPS	QT	° ' "	Latitude, Longitude
	Relative position	Sketch map	QL	m	x, y
Elevation	Absolute position	DGPS	QT	mm to cm	z
		Laser range finder	QT	cm	z
		Altimetry	QT	m to dm	z
Geomorphic setting and environment[3]					
Geomorphology	Cliff top, marine terrace, reef platform, shoreline, beach, etc.	Photography	QL	–	Environmental context
Arrangement	Isolated, clustered, depositional ridge, boulder field, etc.	Mapping	QL	–	Depositional setting
Rugosity	Roughness of reef surface	Surveying	QL	cm to m	Proximity attributes
Morphometric data					
Dimensions	a, b, c axes	Measuring tape, laser range finder	QT	cm	Volume
Shape	Rectangular, ovoid, triangular	Typology	QL	–	Volume
Orientation	Direction of long axis	Compass	QT	Angle (°)	Wave direction
Imbrication	Dip	Clinometer	QT	Angle (°)	Wave direction, transport mode

(continued)

Table 4.1 (continued)

Descriptor	Observation or measurement[4]	Method or tool	Information type[1]	Scale or precision	Interpretation or value
Nature and condition of the boulder[4]					
Lithology	Volcanic and metamorphic rocks	Petrography	QL	Depending on local diversity	Autochthonous or allochthonous
	Sedimentary rocks	Petrography, Palaeontology	QL	Fossil variability	Source area identification
	Coral	Species identification	SQT	Ecological level, elevation above sea level	Source area identification
		Growth direction	QL	Inverted	Overturning during transport
			QL	Normal	Inconclusive
Mobility	Impact scars on boulders and substrate	Percussion marks	QL	–	Transport mode: rolling, saltation
		Striae	QL	–	Transport mode: sliding
Base burying	Depth of burial, surrounding sediment type	Excavation	SQT	Grain size and depth (cm)	Local sediment dynamics
Biological colonisation	Seaweed/lichen cover	Visual aid cover chart	SQT	Surface cover (%)	Immobility or partial mobility
	Inclined fauna/flora bands	Dipping	QT	Angle (°)	Recent mobility
	Fauna/flora zonation	Species identification	QL	Coastal ecological level	Pre-transport source area

(continued)

Table 4.1 (continued)

Descriptor	Observation or measurement	Method or tool	Information type[1]	Scale or precision	Interpretation or value
Denudation	Solution or karstification on limestone	Typology, weathering pit depth	QL or QT	cm	Relative age
	Bioerosion	Lasergrammetry	QT	–	Eroded volume
	Non-carbonate weathering	Schmidt hammer tests	SQT	–	Relative age

Explanation:

[1] *QT* quantitative; *SQT* semi-quantitative; *QL* qualitative; (see Box 6)

[2] Geographical data includes country; location; site; geographical coordinates of the boulder (latitude, longitude, elevation above sea level); precision of GPS coordinates; date of observation

[3] Geomorphic setting includes description of site and relevant additional information such as distance from the shoreline (high-water mark, wave breaking point, cliff edge, reef edge); overall nature of deposit, e.g. whether part of a constructional landform such as a boulder rampart

[4] Boulder condition:

Mobility. Impact scars are valuable indicators of boulder transport mode so it is important to search for and distinguish between these features. Striae indicate sliding of a coral boulder over a reef surface, whereas crushing or percussion marks suggest rolling or saltation motion (Fig. 4.13). Impacts scars are short lived and must be observed within a few months of an event

Weathering stage (relative dating). Boulder fields or ridges are sometimes made up of populations of different ages. Surface colour or roughness might allow discrimination of these populations on the basis of their relative age. For instance, fresh coral boulders observed shortly after emplacement by a HEMI event are whitish in appearance compared to older clasts. *Porites* boulders are smooth when observed within a few years of dislodgment from the living reef but their surface becomes more irregular after decades of weathering

Weathering pit depth. Weathering is strongly related to exposure time, hence differences in solution pit depth might indicate relative age, providing that lithology and exposure conditions are consistent between individual boulders

Base burying. Coastal boulders are part of sedimentary assemblages comprising finer fractions (gravel, sand, mud). At the time of deposition, boulders may be partially or totally embedded in a fine sediment matrix. Post-depositional processes act either to wash away or add to the matrix, depending on the location or time of the year (season). Noting the thickness of loose sediment surrounding a boulder is an indicator of local sediment dynamics, especially where repeated surveys are possible

Life position of coral colonies. If coral colonies in a boulder fabric present an upside-down facies orientation, this suggests boulder overturning. Sliding-type movement can then be dismissed from flow-velocity estimates using hydrodynamic transport equation

Biological belt inclination. An algal belt might be seen around a boulder that was settled near the high-tide mark prior to reworking by a subsequent inundation event. After redeposition, a dipping biological belt suggests a non-sliding mode of transport

Fig. 4.13 a Coral boulder on the south east coast of Lu Dao Island, southern Taiwan. This was the largest boulder observed to have been transported during Typhoon Tembin in August 2012. Fresh impact scars and crush marks over the entire surface indicate that rolling and/or saltation was the mode of transport. The person in the background is pointing downwards at percussion marks on the reef surface; these indicate both the direction of boulder transport and the minimum distance moved (using the impact mark located farthest away). **b–c** Close-up view of chatter marks on the reef surface. (Photos by J. Terry, November 2012)

Box 6
Nature of Boulder Data

QT—Quantitative Dataset
The quantitative dataset comprises *objective* information, i.e. data that are independent of the observer. Size (if using a measuring tape), GPS coordinates and location name are all examples of objective observations.

SQT—Semi-Quantitative Dataset
The semi-quantitative dataset contains measurements that have absolute values, but the precision varies between observers depending on their level of training or experience. Consequently, these data are not totally independent of the observer and measurements yield only an approximation of the actual quantity. For example, the algae or lichen cover (%) on a boulder's surface can be estimated using a visual aid chart, but results might vary slightly from one person to another.

QL—Qualitative Dataset
Qualitative data encompasses boulder information that is subjective or observer-dependent: such as weathering stage, shape, percussion scars, bioerosion zonation and colour (if not determined from a standard colour chart). These data are 'imperfect' and their description will probably vary from one observer to another. Nevertheless they are helpful in a number of ways, for instance with identifying individual boulders or separating new from old deposits in a boulder field, or by revealing precious information on transport mode.

References

Bishop PM, Hughes MG (1989) Imbricate and fitted fabrics in coastal boulder deposits on the Australian east coast. Geology 17:544–547

Bryant EA (2001) Tsunami: the underrated hazard. Cambridge University Press, Stanford, p 350

Bryant EA, Nott JF (2001) Geological indicators of large tsunami in Australia. Nat Hazards 24:231–249. doi:10.1023/A:1012034021063

Bryant EA, Young RW, Price DM (1992) Evidence of tsunami sedimentation on the southeastern coast of Australia. J Geol 100:753–765. doi:10.1086/629626

Chappell J, Omura A, Esat T, McCulloch M, Pandolfi J, Ota Y, Pillans B (1996) Reconciliation of late Quaternary sea levels derived from coral terraces at Huon Peninsula with deep sea oxygen isotope records. Earth Planet Sci Lett 141:227–236. doi:10.1016/0012-821X(96)00062-3

Chen B, Chen Z, Stephenson W, Finlayson B (2011) Morphodynamics of a boulder beach, Putuo Island, SE China coast: the role of storms and typhoon. Mar Geol 283(1–4):106–115. doi:10.1016/j.margeo.2010.10.004

Courtney C (2012) Holocene palaeoenvironmental reconstruction of four coastal sites in southern New South Wales, Australia: implications for the Australian megatsunami hypothesis. Unpublished PhD thesis, University of New South Wales, Australia

Done TJ (1992) Effects of tropical cyclone waves on ecological and geomorphological structures on the Great Barrier Reef. Cont Shelf Res 12(78):859–872. doi:10.1016/0278-4343(92)90048-O

Engel M, Brückner H, Wennrich V, Scheffers A, Kelletat D, Vött A, Schäbitz F, Daut G, Willershäuser T, May SM (2010) Coastal stratigraphies of eastern Bonaire (Netherlands Antilles): New insights into the palaeo-tsunami history of the southern Caribbean. Sed Geol 231:14–30. doi:10.1016/j.sedgeo.2010.08.002

Engel M, May SM (2012) Bonaire's boulder fields revisited: evidence for Holocene tsunami impact on the Leeward Antilles. Quatern Sci Rev 54:126–141. doi:10.1016/j.quascirev.2011.12.011

Etienne S (2010) Les crises géomorphologiques. Impacts sur les paysages naturels, impacts sur les sociétés contemporaines. Unpubl Hab thesis, Univ Clermont-Ferrand, p 389

Etienne S, Paris R (2010) Boulder accumulations related to storms on the south coast of the Reykjanes Peninsula (Iceland). Geomorphology 114(1–2):55–70. doi:10.1016/j.geomorph.2009.02.008

Etienne S, Buckley M, Paris R, Nandasena AK, Clark K, Chagué-Goff C, Goff J, Richmond B (2011a) The use of boulders for characterizing past tsunamis: lessons from the 2004 Indian Ocean and 2009 South Pacific tsunamis. Earth Sci Rev 107:75–90. doi:10.1016/j.earscirev.2010.12.006

Etienne S, Paris R, Switzer A, Terry JP (2011b) The GEOBOULDER portal: a coastal boulder survey tool to improve modelling of sediment transport during high-energy events. 22nd Pacific Science Congress of the Pacific Science Association (PSA), 14–18 June, Kuala Lumpur, Malaysia

Etienne S, Terry JP (2012) Coral blocks, gravel tongues and sand sheets: features of coastal accretion and sediment nourishment by Cyclone Tomas (March 2010) on Taveuni Island, Fiji. Geomorphology 175–176:54–65. doi:10.1016/j.geomorph.2012.06.018

Felton EA, Crook KAW (2003) Evaluating the impacts of huge waves on rocky shorelines: an essay review of the book 'Tsunami—the underrated hazard'. Mar Geol 197:1–12. doi:10.1016/S0025-3227(03)00086-0

Fichaut B, Suanez S (2011) Quarrying, transport and deposition of cliff-top storm deposits during extreme events: Banneg Island Brittany. Mar Geol 283(1–4):36–55. doi:10.1016/j.margeo.2010.11.003

Frohlich C, Hornbach MJ, Taylor FW (2011) Megablocks. In: Hopley D (ed) Encyclopedia of modern coral reefs: structure, form and process. Springer Netherlands, Dordrecht, pp 679–683

Gienko G, Terry JP (forthcoming) 3D modelling of boulder shape and volume using digital photogrammetric techniques. Applications in coastal geomorphology. International Association of Geomorphologists 8th International Conference on Geomorphology, 27–31 August 2013, Paris, France

Goff J, Dominey-Howes D (2009) Australian palaeotsunamis—do Australia and New Zealand have a shared trans-Tasman prehistory? Earth-Sci Rev 97:147–154. doi:10.1016/j.earscirev.2009.09.003

Goto K, Chavanich SA, Imamura F, Kunthasap P, Matsui T, Minoura K, Sugawara D, Yanagisawa H (2007) Distribution, origin and transport process of boulders deposited by the 2004 Indian Ocean tsunami at Pakarang Cape, Thailand. Sed Geol 202(4):821–837. doi:10.1016/j.sedgeo.2007.09.004

Goto K, Shinozaki T, Minoura K, Okada K, Sugawara D, Imamura F (2010a) Distribution of boulders at Miyara Bay of Ishigaki Island, Japan: a flow characteristic indicator of tsunami and storm waves. Island Arc 19(3):412–426. doi:10.1111/j.1440-1738.2010.00721.x

Goto K, Kawana T, Imamura F (2010b) Historical and geological evidence of boulders deposited by tsunamis, southern Ryukyu Islands Japan. Earth Sci Rev 102(1–2):77–99. doi:10.1016/j.earscirev.2010.06.005

Goto K, Miyagi K, Kawamata H, Imamura F (2010c) Discrimination of boulders deposited by tsunamis and storm waves at Ishigaki Island, Japan. Mar Geol 269:34–45. doi:10.1016/j.margeo.2009.12.004

Goto K, Sugawara D, Ikema S, Miyagi T (2012) Sedimentary processes associated with sand and boulder deposits formed by the 2011 Tōhoku-oki tsunami at Sabusawa Island, Japan. Sed Geol 282:188–198. doi:10.1016/j.sedgeo.2012.03.017

Hayne M, Chappell J (2001) Cyclone frequency during the last 5000 years at Curacoa Island, north Queensland, Australia. Palaeogeogr Palaeoclimatol Palaeoecol 168(3–4):207–219. doi:10.1016/S0031-0182(00)00217-0

Hills ES (1970) Fitting, fretting and imprisoned boulders. Nature 226:345–347. doi:10.1038/226345b0

Hobbs WH (1933) Reviews (Reviewed work: Coral Reefs and Atolls by J.S. Gardiner). J Geol 41(2):219–222

Kázmér M, Taboroši D (2012) Bioerosion on the small scale- examples from the tropical and subtropical littoral. Hantkeniana 7:37–94

Kelletat D, Scheffers SR, Scheffers A (2007) Field signatures of the SE-Asian mega-tsunami along the west coast of Thailand compared to Holocene paleo-tsunami from the Atlantic region. Pure Appl Geophys 164(2–3):413–431. doi:10.1007/s00024-006-0171-6

Kogure T, Matsukura Y (2010) Instability of coral limestone cliffs due to extreme waves. Earth Surf Proc Land 35:1357–1367. doi:10.1002/esp.2046

Kortekaas S, Dawson A (2007) Distinguishing tsunami and storm deposits: An example from Martinhal SW Portugal. Sed Geol 200(3–4):208–221. doi:10.1016/j.sedgeo.2007.01.004

Loevenbruck A, Hebert H, Schindele F, Sladen A, Lavigne F, Brunstein D, Wassmer P, Paris R (2007) Detailed modeling of the 2004 tsunami flooding in the Banda Aceh and Lhok Nga districts (Sumatra, Indonesia). American Geophysical Union, Fall Meeting 2007, abstract #S53A-1042

Lorang MS (2011) A wave-competence approach to distinguish between boulder and mega-clast deposits due to storm waves versus tsunamis. Mar Geol 283(1–4):90–97. doi:10.1016/j.margeo.2010.10.005

McKenna J, Jackson DWT, Cooper JAG (2011) In situ exhumation from bedrock of large rounded boulders at the Giant's Causeway, Northern Ireland: An alternative genesis for large shore boulders (mega-clasts). Mar Geol 283(1–4):25–35. doi:10.1016/j.margeo.2010.09.005

Medina F, Mhammdi N, Chiguer A, Akil M, Jaaidi EB (2011) The Rabat and Larache boulder fields; new examples of high-energy deposits related to storms and tsunami waves in north-western Morocco. Nat Hazards 59:725–747. doi:10.1007/s11069-011-9792-x

Morton RA, Richmond BM, Jaffe BE, Gelfenbaum G (2006) Reconnaissance investigation of Caribbean extreme wave deposits- preliminary investigations, interpretations, and research directions. Open-File Report 2006-1293, USGS

Neef G, Hendy C (1988) Late Pleistocene-Holocene acceleration of uplift rate in southwest Erromango Island, southern Vanuatu, South Pacific: relation to the growth of the Vanuatuan mid sedimentary basin. J Geol 96(4):481–494. doi:10.1086/629242

Noormets R, Felton EA, Crook KAW (2002) Sedimentology of rocky shorelines: 2. Shoreline megaclasts on the north shore of Oahu, Hawaii-origins and history. Sed Geol 150(1–2):31–45. doi:10.1016/S0037-0738(01)00266-4

Noormets R, Crook KAW, Felton EA (2004) Sedimentology of rocky shorelines: 3. Hydrodynamics of megaclast emplacement and transport on a shore platform, Oahu, Hawaii. Sediment Geol 172:41–65. doi:10.1016/j.sedgeo.2004.07.006

Nott J (1997) Extremely high-energy wave deposits inside the Great Barrier Reef, Australia: determining the cause-tsunami or tropical cyclone. Mar Geol 41(1–4):193–207. doi:10.1016/S0025-3227(97)00063-7

Nott J (2000) Records of prehistorical tsunamis from boulder deposits evidence from Australia. Sci Tsunami Hazards 18(1):3–14

Nott J (2003) Waves, coastal boulder deposits and the importance of the pre-transport setting. Earth Planet Sci Lett 210(1–2):269–276. doi:10.1016/S0012-821X(03)00104-3

Paris R, Lavigne F, Wassmer P, Sartohadi J (2007) Coastal sedimentation associated with the December 26, 2004 tsunami in Lhok Nga, west Banda Aceh (Sumatra, Indonesia). Mar Geol 238(1–4):93–106. doi:10.1016/j.margeo.2006.12.009

Paris R, Wassmer P, Sartohadi J, Lavigne F, Barthomeuf B, Desgages É, Grancher D, Baumert P, Vautier F, Brunstein D, Gomez C (2009) Tsunamis as geomorphic crisis: lessons from the December 26, 2004 tsunami in Lhok Nga, west Banda Aceh (Sumatra, Indonesia). Geomorphology 104:59–72. doi:10.1016/j.geomorph.2008.05.040

Paris R, Fournier J, Poizot E, Etienne S, Morin J, Lavigne F, Wassmer P (2010) Boulder and fine sediment transport and deposition by the 2004 tsunami in Lhok Nga (western Banda Aceh, Sumatra, Indonesia): a coupled offshore—onshore model. Mar Geol 268:43–54. doi:10.1016/j.margeo.2009.10.011

Paris R, Naylor LA, Stephenson WJ (2011) Boulders as a signature of storms on rock coasts. Mar Geol 283:1–11. doi:10.1016/j.margeo.2011.03.016

Ricard M (1985) Les récifs coralliens de Polynésie française. Atoll de Rangiroa, Archipel des Tuamotu. Proceedings of Fifth International Coral Reef Congress, Tahiti, 1:161–193

Richmond BM, Buckley M, Etienne S, Chagué-Goff C, Clark K, Goff J, Dominey-Howes D, Strotz L (2011) Deposits, flow characteristics, and landscape change resulting from the September 2009 South Pacific tsunami in the Samoan Islands. Earth Sci Rev 107(1–2):38–51. doi:10.1016/j.earscirev.2011.03.008

Rowe DAC, Khan SA, Robinson E (2009) Hurricanes or tsunami? Comparative analysis of extensive boulder arrays along the southwest and north coasts of Jamaica: lessons for coastal management. In: Macgregor M, Dodman D, Barker D (eds) Global change and Caribbean vulnerability: environment, economy and society at risk? UWI Press, Kingston

Saintilan N, Rogers K (2005) Recent storm boulder deposits on the Beecroft Peninsula, New South Wales. Aust Geogr Res 43:429–432. doi:10.1111/j.1745-5871.2005.00344.x

Scheffers A, Kelletat D (2003) Sedimentologic and geomorphologic tsunami imprints world-wide—a review. Earth Sci Rev 63(1–2):83–92. doi:10.1016/S0012-8252(03)00018-7

Scheffers A, Scheffers S (2007) Tsunami deposits on the coastline of west Crete (Greece). Earth Planet Sci Lett 259(3–4):613–624. doi:10.1016/j.epsl.2007.05.041

Scheffers SR, Haviser J, Browne T, Scheffers A (2009) Tsunamis, hurricanes, the demise of coral reefs and shifts in prehistoric human populations in the Caribbean. Quatern Int 195(1–2):69–87. doi:10.1016/j.quaint.2008.07.016

Scoffin TP (1993) The geological effects of hurricanes on coral reefs and the interpretation of storm deposits. Coral Reefs 12:203–221. doi:10.1007/BF00334480

Spiske M, Böröcz Z, Bahlburg H (2008) The role of porosity in discriminating between tsunami and hurricane emplacement of boulders- a case study from the Lesser Antilles, southern Caribbean. Earth Planet Sci Lett 268:284–396. doi:10.1016/j.epsl.2008.01.030

Stephenson WJ, Naylor LA (2011) Geological controls on boulder production in a rock coast setting: Insights from South Wales UK. Mar Geol 283(1–4):12–24. doi:10.1016/j.margeo.2010.07.001

Switzer AD, Burston JM (2010) Competing mechanisms for boulder deposition on the southeast Australian coast. Geomorphology 114:42–54. doi:10.1016/j.geomorph.2009.02.009

Tappin DR, Evans HM, Jordan CJ, Richmond B, Sugawara D, Goto K (2012) Coastal changes in the Sendai area from the impact of the 2011 Tōhoku-oki tsunami: Interpretations of time series satellite images, helicopter-borne video footage and field observations. Sed Geol 282:151–174. doi:10.1016/j.sedgeo.2012.09.011

Terry JP (2004) Geomorphic features of Niue Island: chasms, caves and other karst varieties. In: Terry JP, Murray WE (eds) Niue Island, geographical perspectives on the Rock of Polynesia. International Scientific Council for Island Development, UNESCO, Paris, pp 75–88

Terry JP, Nunn PD (2003) Interpreting features of carbonate geomorphology on Niue Island, a raised coral atoll. Zeitschrift für Geomorphologie, Supplement Band 131:43–57

Terry JP, Etienne S (2010) "Stones from the dangerous winds": reef platform mega-clasts in the tropical Pacific Islands. Nat Hazards 56(3):567–569. doi:10.1007/s11069-010-9697-0

Terry JP, Etienne S, Paris R, Switzer A (2011) Introducing GEOBOULDER—an open-access web repository of coastal boulder data to facilitate geomorphic analysis of high-energy sediment transport events. Asia Oceania Geosciences Society (AOGS), 8th Annual Conference, 8–12 August 2011, Taipei, Taiwan

Trudgill ST (1976) The marine erosion of limestone on Aldabra Atoll, Indian Ocean. Zeitschrift für Geomorphologie, Supplementband 26:164–200

Trudgill S (2011) Solution processes/reef erosion. In: Hopley D (ed) Encyclopedia of modern coral reefs: structure, form and process. Springer, Dordrecht, pp 1024–1027

Watt SG, Jaffe BE, Morton RA, Richmond BM, Gelfenbaum G (2010) Description of extreme-wave deposits on the northern coast of Bonaire, Netherlands Antilles. USGS Open-File Report 2010-1180

Yu K, Zhao J, Collerson KD, Shi Q, Chen T, Wang P, Liu T (2004) Storm cycles in the last millennium recorded in Yongshu Reef, southern South China Sea. Palaeogeogr Palaeoclimatol Palaeoecol 210(1):89–100. doi:10.1016/j.palaeo.2004.04.002

Yu K, Zhao J, Shi Q, Meng Q (2009) Reconstruction of storm/tsunami records over the last 4000 years using transported coral blocks and lagoon sediments in the southern South China Sea. Quatern Int 195(1–2):128–137. doi:10.1016/j.quaint.2008.05.004

Chapter 5
Case Study: Coral Boulder Fields on Taveuni Island Coasts, Fiji

Abstract The coastline of Taveuni Island in the southwest Pacific was struck by a category-4 tropical cyclone in March 2010. Post-storm field investigations of the coastal geomorphic impacts concentrated on an area in central Taveuni where protected fringing reefs and coastlines form part of the Bouma National Heritage Park. Here, a range of cyclone constructional imprints were found to have supplemented existing coastal sediments. Fresh coral boulders strewn across reef platforms indicate that TC Tomas had sufficient power to deliver new coral blocks, but that this material comprises a relatively minor component of pre-existing boulder fields. Comparison between the dimensions of fresh and older blocks reveals that unknown earlier events (storms or tsunamis) produced much larger debris and therefore presumably generated more energetic flow velocities across the fringing reefs than TC Tomas did. Analysis of calcarenite slabs quarried from in situ beachrock exposures was also particularly useful for calculating surging flow velocities at the shoreline.

5.1 Introduction and Aims

The case study presented in this chapter forms a component of continuing investigations on coastal geomorphic change experienced on Taveuni Island in northern Fiji, especially the effects of Tropical Cyclone (TC) Tomas in March 2010. The summary here focuses on the characteristics of reef-platform boulder fields and beachrock excavation in four locations. Readers are directed to Etienne and Terry (2012) for a more comprehensive account dealing with a wider range of coastal features than those described here.

TC Tomas in mid-March 2010 was a category-4 intensity cyclone that affected the northern and eastern islands of Fiji in the southwest Pacific. The eastern side of Taveuni Island in particular experienced monstrous waves and powerful storm surge and as the eye passed within 30 km of the coast. Impacts on coastal geomorphology were investigated within a few months of the cyclone on Taveuni. Clear geomorphic and sedimentary signatures were left by TC Tomas in Taveuni's eastern coastal landscape, distinguishable as both quarrying and scouring features

J. P. Terry et al., *Reef-Platform Coral Boulders*, SpringerBriefs in Earth Sciences,
DOI: 10.1007/978-981-4451-33-8_5, © The Author(s) 2013

related to the production, removal and/or remobilisation of material, and associated depositional assemblages occurring as various types of sedimentary structures. Amongst the principal aims were to identify the geomorphic imprints resulting from this major storm event, to examine spatial variations in storm effects along the coastline, and to extract quantitative information on the grain-size distributions of coarse deposits, including fresh reef-platform coral boulders (RPCBs) and fragments of beachrock excavated from known sources. The physical characteristics of these deposits were then subject to numerical modelling in order to determine flow velocities and sediment transport capability of the storm-wave conditions prevailing during TC Tomas. This further allows comparison with similar but more substantial reef-top deposits laid down by unknown high-energy marine inundation (HEMI) events at earlier times.

5.2 Background to Study Area

Taveuni Island is an elongated volcanic island in the north of the Fiji archipelago, traversed by the 180° meridian. Approximately 42 km long and 10–14 km wide, the island is oriented in a north east—south west direction. Although Taveuni's basaltic shield volcanoes are inactive, the topography is dominated by an impressive series volcanic peaks forming a highland spine along the centre of the island; several peaks reach elevations over 1000 m a.s.l (Fig. 5.1). Facing the moist south

Fig. 5.1 Taveuni Island in Northern Fiji. Most of the coastline and coral reefs in the boxed area on the map have been established as a Marine Protected Area. The satellite image (courtesy of Google Earth, GeoEye) shows the location of coral boulder fields on fringing reefs and displaced fragments of beachrock that were examined on the central east coast between Bouma and Lavena villages

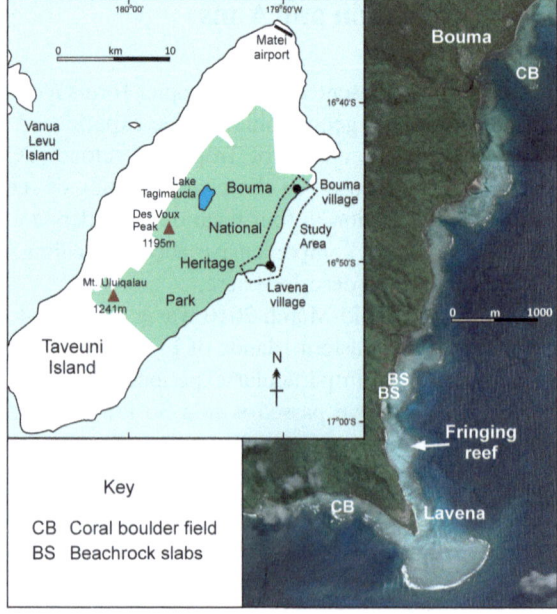

east trade winds means that the south east coast receives annual rainfall of approximately 5,000 mm, whilst precipitation near the summit of the interior mountains is almost double that figure, approaching 10,000 mm/year. Owing to the year-round very wet climate, the island's rugged landscape is deeply dissected by steep river valleys and thickly vegetated with tropical rainforest (Ash 1987). Taveuni is therefore fondly called the 'Garden Island of Fiji'.

Much of Taveuni's coastline is pristine and undisturbed. The attractive landscapes and rich biodiversity of the island, both in terms of terrestrial and marine life, have enabled small scale ecotourism ventures to grow and become an important component of the local economy. Although coral reefs do not encircle the island entirely, they are well developed along the northern and eastern central coasts, extending as 100–500 m broad platforms from the shoreline. To safeguard these reefs and adjacent shorelines, a Marine Protected Area (MPA) was established in 1998. The MPA lies within the greater Bouma National Heritage Park (BNHP 2011). At Lavena and Bouma villages (Fig. 5.1), locally-managed visitor centres organise eco-adventure activities that attract a daily flow of international tourists vacationing at resorts elsewhere on the island. Fees payable provide a steady cash income that benefits the rural population, who otherwise pursue traditional Fijian village-based subsistence lifestyles and are the proud custodians of their unspoilt natural heritage. Consequently, tropical cyclones that cause damage to coral reefs and coastlines are perceived as 'negative' events by local communities, so the TC Tomas event provided additional impetus for this study within this context.

5.3 Features of Tropical Cyclone Tomas, March 2010

TC Tomas (12–17 March 2010) was the ninth cyclone to develop within the South Pacific basin over the 2009/2010 wet season (November to April). The latter part of this season was marked by unusual climatic activity east of 180°, where four cyclones reached category-3 or 4[1] intensity in rapid succession over a relatively brief period from February to March 2010 (Terry and Etienne 2010a, b). Of these storms, TC Tomas was the most destructive for Fiji in almost a decade (Terry et al. 2004). El Niño conditions of moderate strength played a part in the heightened levels of disturbance in early 2010 (FMS 2010), allowing warm sea temperatures to stretch far east of the Date Line. This warm pool developed into a broad region of low pressure within which atmospheric instability gave rise to numerous eddies showing potential for spin up into cyclonic vortices.

The nascent depression that later matured into TC Tomas first became organised north east of the Fiji Islands in Samoan waters. Initially following a west south-westwards track, the storm strengthened and attained cyclone status (winds >34 kts)

---

[1] Australian intensity scale for tropical cyclones in the south west Pacific, which is used by both the Australian Bureau of Meteorology (BoM) and the Fiji Meteorological Service (FMS).

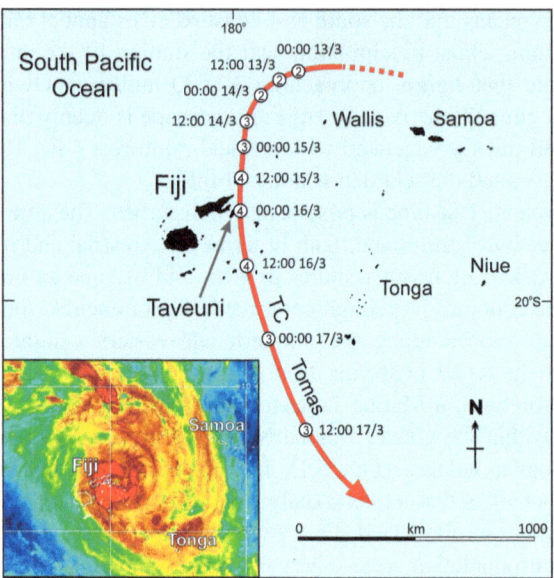

Fig. 5.2 Path of Tropical Cyclone Tomas in mid-March 2010 as the system traversed eastern Fiji waters. Positions of the cyclone eye at six-hourly intervals are shown by circular markers, with adjacent timings given in Fiji Standard Time (FST, 12 h ahead of Greenwich Mean Time). The intensity categories written in the eye markers correspond to the following wind speeds (sustained maximums over 10 min intervals): category-1: 34–47 knots; category-2: 48–63 knots; category-3: 64–85 knots; category-4: 86–107 knots. *Inset figure* at bottom left shows a satellite thermal image of TC Tomas at 12:30 a.m. FST on 16 March. The storm eye is located near the northern tip of Taveuni Island while at category-4 intensity. Original satellite image courtesy of NOAA

north of Wallis Island at 12 p.m. Fiji Standard Time[2] (midnight GMT) on 12 March. Thereafter, the system matured rapidly. It decelerated in speed but intensified to category-4 strength, as illustrated in Fig. 5.2. The eye of TC Tomas passed within 30 km of the east coast of Taveuni Island in northern Fiji on 15 March and then migrated through the Lau group of islands over the next day. The significance of this track for the Taveuni coastline is that sustained forceful winds of 100 knots[3] (185 km/hr) and gusts up to 140 knots (259 km/hr) were blowing onshore as the storm approached. The furious winds drove large waves, heavy swells and powerful storm surge ashore, causing severe flooding of low-lying coastal areas and much damage to coastal vegetation, villages and transport networks.

---

[2] Fiji Standard Time (FST) is 12 h ahead of Greenwich Mean Time (GMT). All subsequent times and dates are FST.

[3] Wind speeds refer to sustained winds over 10 min averaging times.

5.4 Field Procedures and Observations

Four months after TC Tomas, detailed geomorphic field investigations were carried out in July 2010 along the eastern coast of Taveuni Island, with a deliberate concentration of effort between Bouma and Lavena villages. This section of coastline experienced the full brunt of storm-driven waves due to its exposure in relation to the position and orientation of the cyclone track. Table 5.1 provides specific locations and details of bouldery coastal deposits examined at four field sites.

Coral boulder fields in two in reef-platform areas were mapped using a GPS Garmin 76 CSX device to 2–3 m precision. The dimensions of individual boulders (*a*, *b*, *c* axis length) were concurrently measured and recorded. During field work, a number of Fijian villagers with an intimate knowledge of the local coastline acted as guides. A principal task was to identify which reef-platform boulders were fresh deposits that had appeared in the aftermath of TC Tomas. This was easily accomplished through recognition by the guides and the noticeable greenish-white colouration of new boulders, which stood out clearly against their older weathered neighbours that pre-dated the cyclone in question (Fig. 5.3). The fresh clasts were dispersed across the reef platform and thus augmented the existing boulder fields. Of interest was that no rubble ramparts of mixed coralline gravels and boulders were observed parallel to reef crests, such as of the type described both in Tuvalu and Samoa after violent cyclones struck those neighbouring countries in 1972 and 1990 respectively (Maragos et al. 1973; Zann 1991). Instead, however, three tongues of loose coral gravel were formed on Taveuni's reef flats, oriented perpendicular to the reef edge. Several storm ridges of shingle were also built up along the highest part of beaches between Bouma and Lavena. Several villagers independently confirmed these as new constructional features resulting from TC Tomas (see Etienne and Terry 2012, for further description).

A total of 196 coral boulders were observed, of which 38 were fresh boulders produced by TC Tomas, and 158 were old boulders. According to the methods outlined by Etienne and Paris (2010), any scars, gouges, percussion marks or striations on boulder surfaces, or on the adjacent reef platform, were recorded as indicators

Table 5.1 Details of coastal measurement sites in eastern Taveuni: coral boulders and beachrock fragments

Deposit type	Coastal site	Location	Number of clasts examined
Coral boulder field (Bouma)	Reef platform	S16°49.507′ W179°51.892′	51 boulders
Coral boulder field (Lavena)	Reef platform	S16°52.376′ W179°53.266′	37 boulders
Beachrock fragments	Shore and back-beach locations	S16°51.552′ W179°52.905′	19 clasts
Beachrock fragments	Shore and back-beach locations	S16°51.500′ W179°52.888′	31 clasts

Fig. 5.3 a Photograph of the fringing reef near Lavena village showing several of the numerous coral boulders strewn across the reef surface. The boulder with the greenish-white colour nearest the observer is a new clast thrown up by TC Tomas. It contrasts markedly with the surrounding group of older boulders behind, brown and weathered in appearance, which pre-date this cyclone. **b** Angular beachrock fragments excavated from in situ exposures along the Bouma–Lavena shoreline. (Photos by J. Terry, July 2010)

of the mode of boulder transport. Nevertheless, the majority of pre-existing coral boulders bore no indication of recent movement. The lack of surficial scorings on them or on the reef flat in their immediate vicinity is interpreted as evidence that TC Tomas did not generate sufficient wave energy to set them in motion.

In several places, exposed beachrock at the shoreline provided a source of coarse clastic debris. Angular slab-shaped beachrock fragments had been broken off and subsequently moved. Using similar procedures to those mentioned above, two groups of clasts were inspected approximately 1 km north of Lavena (Fig. 5.3). Fifty clasts in total were measured, all transported by TC Tomas. Careful scrutiny revealed that beachrock-derived material had been carried in two directions, either pushed inland by wave runup or pulled backwards by return flow towards the beach–reef interface and onto the adjacent landward edge of the fringing reef.

5.5 Results and Discussion

5.5.1 Boulder Quarrying and Remobilisation

Expansive coral boulder fields occupy the fringing reefs in two areas near Bouma and Lavena villages. On many individual boulders, however, features of bioerosion (cavities eroded by sea urchins and other marine creatures), solution weathering (karstification) and marine erosion (nascent wave-cut notches) revealed that the majority had stood in place for a considerable period of time prior to TC Tomas. Among the total number of boulders inspected (n = 196), the recent deposits (n = 38) were easily distinguished by their outer greenish-white colouration. All but two of these are relatively small boulders (<0.5 m³), with the most sizeable fresh boulder having a volume of 4.85 m³. Information on boulder volume frequencies for both fresh and older deposits is displayed in Fig. 5.4. Perhaps unexpectedly, in as much as the boulder fields did contain some fresh deposits, there was a general scarcity of recently uprooted or overturned coral colonies on the reef flat itself. Such evidence suggests that coral colonies on the reef edge or external reef slopes

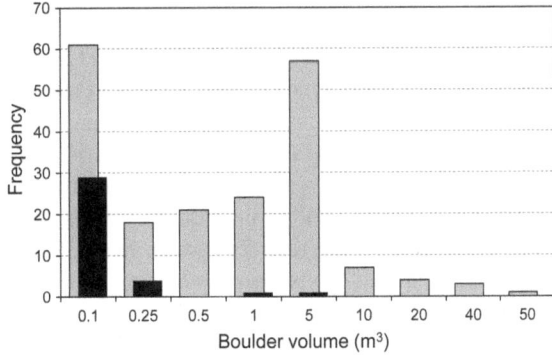

Fig. 5.4 Frequency distribution of coral boulders by volume on Taveuni reef platforms near Bouma and Lavena, as derived from measurements of boulder axial dimensions (a,b,c axis). Fresh boulders emplaced by TC Tomas are represented by solid black bars, while grey bars include all boulders measured (n = 196)

contributed primarily to the production of new boulders, although sea conditions during the field expedition were unsuitable for submarine work to confirm this.

It is noteworthy that pre-existing coral boulders are substantially bigger than those delivered by TC Tomas: the largest exceeds 40 m^3 in size, while 16 others are bigger than the largest boulder moved during the storm. No evidence of recent mobilisation could be found on these old boulders: neither scars nor crush marks on the boulder faces (which would indicate rolling), nor grooves on the adjacent reef surfaces (which would indicate sliding). From this, the following conclusions may be drawn. First, the boulder fields were mostly constructed prior to TC Tomas by unknown HEMI events, probably of significant antiquity. Second, waves driven onshore by TC Tomas were unable to mobilise the older boulders. Third, TC Tomas added approximately 20 % to the sum of boulders comprising the existing deposits.

At the shoreline, the quarrying and transport of fresh beachrock fragments is of special interest. Examining beachrock fragments offers one key benefit over RPCBs: identification of their original source is often immediately possible, simply from their neat 'jigsaw fit' into nearby exposures of in situ beachrock. Knowing the exact source point enables the displacement of beachrock fragments by cyclonic waves to be determined precisely. Subsequently, measuring fragment dimensions and transport distance allows reconstruction of both runup and backwash flow velocities at the shoreline. Such information is valuable in another way: it reveals the available energy at the shoreline after waves have been attenuated by passing over the fringing reef. Measured clasts quarried from beachrock near Lavena are very heterogeneous in size: the smallest weighs 11 kg whereas the biggest is over 400 kg. Although 90 % of the boulders weigh less than 150 kg, this can be explained by the jointed nature of the beachrock material, with calcarenite slabs being easily broken in smaller pieces during wave transport. Hydrodynamic data deduced from beachrock clasts can be compared with wave energy on the reef platform itself, obtained from similar calculations using values derived from the reef-top boulders mentioned above.

5.5.2 *Transport Mechanisms and Flow Velocities*

During post-fieldwork data treatment, the minimum flow velocities needed to set in motion coral boulders of particular shapes and in various pre-transport settings were estimated using the hydrodynamic transport equations published by Nott (2003) and revised by Nandasena et al. (2011). Owing to a lack of evidence for significant erosion on the reef surface, the assumption is made that the fresh boulders were quarried from submerged reef-crest or fore-reef locations. Using hydrodynamic equations therefore yields minimum velocities for the flows that emplaced these boulders torn from the reef framework by TC Tomas. Furthermore, when applied to pre-existing boulders showing no signs of movement (assumed partly-subaerial environment on the reef platform), the equations give an indication of flow velocities that were not achieved by the strongest waves driven over the reef by the cyclone.

Simultaneous mapping of combined data from fresh and older boulders then becomes valuable, as this provides a first approximation of spatial patterns in flow velocity ranges experienced during TC Tomas over the reef platform (Fig. 5.5).

Following on from this, a transport histogram plotted for the entire dataset indicates the minimum flow velocity values required to move coral boulders according to the various possible modes of boulder transport, i.e. sliding, rolling or lifting (Fig. 5.6). By adopting the conservative approach that fresh coral boulders were transported on the reef surface only by sliding or rolling, i.e. without saltation, minimum flow velocities lie between 1 and 3.8 m/s. This reveals that in spite of TC Tomas attaining category-4 intensity, waves breaking across the fringing reefs generated insufficient flow velocity to set in motion many pre-existing boulders,

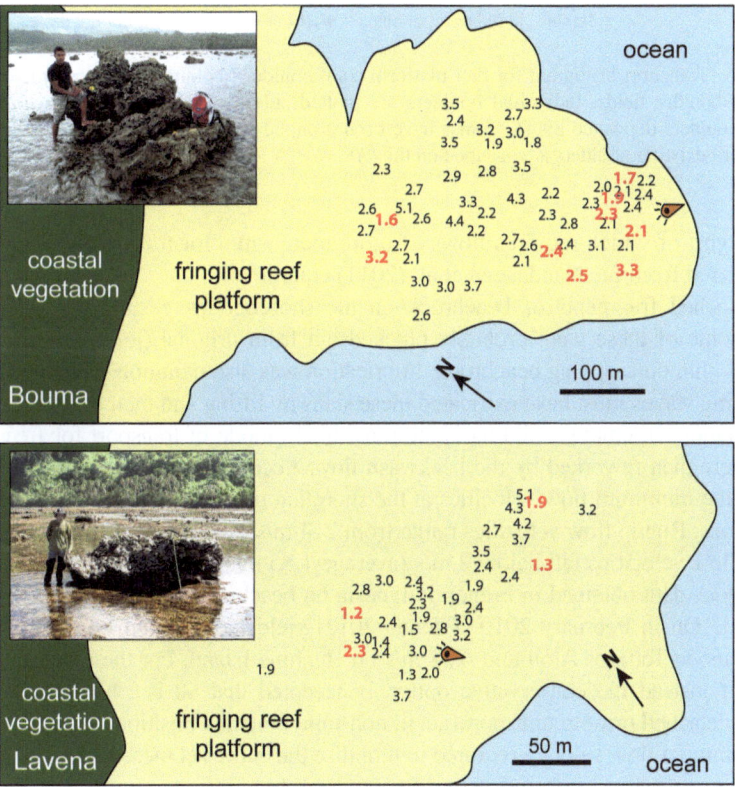

Fig. 5.5 Minimum water flow velocities (m/s) needed to set in motion coral boulders resting on the fringing reef near Bouma (*above*) and Lavena (*below*), as deduced from hydrodynamic equations (Nandasena et al. 2011) applied to measured boulder dimensions. See Fig. 5.1 for the location of the two boulder fields. Red values refer to fresh coral boulders produced by TC Tomas wave action. Black values represent pre-existing boulders that were not displaced during the storm. Together, the two sets of values give an appreciation of the minimum (red) and maximum (black) flows experienced from place to place across the reef flat during TC Tomas. The eye symbols show the direction of view for the inset photographs. (Photos by J. Terry, July 2010)

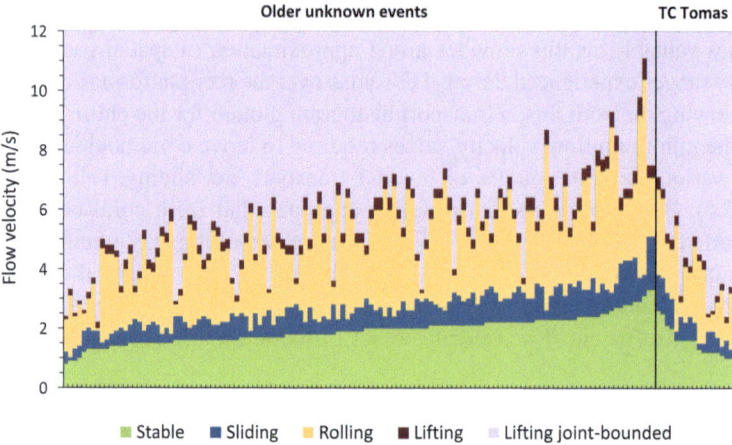

Fig. 5.6 Transport histogram for reef-platform coral boulders: combined data from Bouma and Lavena boulder fields. Individual boulders are plotted side by side along the horizontal axis. Fresh boulders deposited by TC Tomas have been grouped on the *right*-hand side of the graph, while pre-existing boulders are grouped on the *left*

the largest of which requires flows of more than 5 m/s for transportation (sliding) as inferred from the Nandasena et al. (2011) equations.

Detached fragments of beachrock on the shoreline were similarly helpful. At least some of these were probably plucked out from original (joint-bounded) positions within outcropping beachrock. Imbrication was also common. This implies that incoming waves must have extricated these slabs by lifting and then transported them by saltation. In contrast, sliding is the most likely mode of transport for freed slabs that were then reworked by the backwash flow. Accordingly, it is possible to reconstruct the minimum flow velocities at the shoreline using the familiar hydrodynamic equations. Runup flow velocities range from 2–4 m/s (average 3.41 m/s) while backwash flow velocities fall below 2 m/s (average 1.85 m/s) (Fig. 5.7). For comparison, analogous data obtained in French Polynesia on beachrock dislodgment by Tropical Cyclone Oli in February 2010 (Etienne 2012) yield mean runup flow velocities of 3.98 m/s on Tetiaroa Atoll and 4.24 m/s on Huahine Island. For the present Taveuni case, if instead the conservative option is accepted that all beachrock clasts were freely detached prior to transport (i.e. in non-joint bounded positions), then 2 m/s was the minimum flow velocity required to mobilise the entire set measured (Table 5.2).

Flow velocities inferred from fresh coral boulders on reef platforms and beachrock fragments at the shoreline are complementary datasets. Comparison between them provides a valuable indicator of flow velocity reduction over the reef flat due to energy dissipation. Here it can be seen that flow velocities generated by waves breaking at the shoreline during TC Tomas are comparable to, if not greater than, flow velocities over the reef flat. This finding is important as it is somewhat counterintuitive and implies that fringing reefs do not necessarily afford a high degree of protection to the shoreline against cyclone-generated waves.

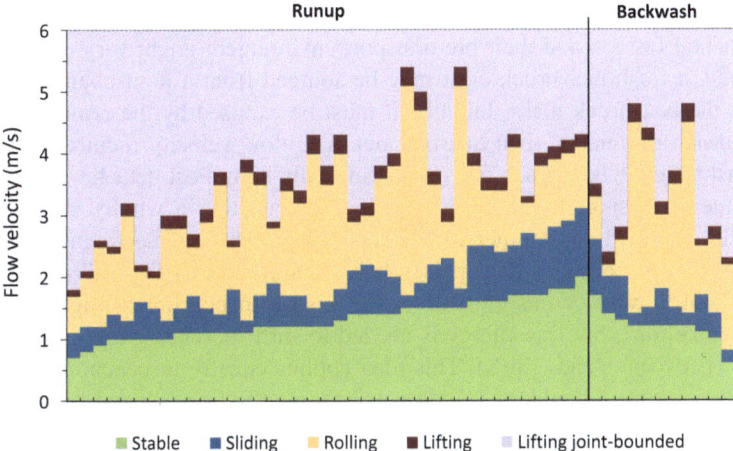

Fig. 5.7 Transport histogram for beachrock fragments at two adjacent shoreline sites, 1 km north of Lavena. Refer to Fig. 5.1 for location of field sites. The vertical axis shows the minimum velocity required to set a clast in motion according to mode of transport. It can be seen that 46 of the 50 slabs are stable with a flow velocity below 1 m/s. When the flow velocity reaches 2 m/s, all beachrock fragments except one are mobilised (see Table 5.2)

Table 5.2 Progressive instability of beachrock fragments with increasing flow velocity (Lavena study site; n = 50)

Flow velocity (m/s)	Stable (%)	Sliding (%)	Rolling (%)	Lifting (%)	Total in motion (%)
1	80	18	2	0	20
1.5	20	44	30	0	74
2	0	30	64	6	100
2.5	0	12	66	22	100
3	0	2	54	44	100
3.5	0	0	34	66	100
4	0	0	16	84	100
4.5	0	0	10	90	100
5	0	0	4	96	100

5.5.3 Caveats to Findings

Calculated flow-velocity values assume that the reef platform is horizontal and smooth. In reality, reef surfaces are rough and uneven owing to surface rugosity associated with the patchwork growth of coral colonies such as microatolls in the back-reef environment. Therefore, higher flow values were probably experienced during TC Tomas.

When attempting to reconstruct boulder transport conditions during a coastal inundation event, knowledge of pre-transport settings is critical for selecting the most appropriate hydrodynamic equation from those available (Nott 2003). One reason is because a tropical cyclone typically generates numerous waves and surges over several hours. Correspondingly, coarse sediments are moved in an incremental fashion and their pre-transport environment might vary over time. To illustrate, a fresh beachrock clast may be sourced from a joint-bounded position within the beachrock mass. Initially, it must be exposed by the removal of overlying beach sediments, then quarried out. The flow velocity required to liberate the clast from the beachrock outcrop is especially high. Post-detachment, however, subsequent transport requires lower velocities in either a wholly submerged or partially subaerial environment. This depends on whether the fragment is transferred into a beach, reef flat or back-beach setting. Here, Fig. 5.7 demonstrates how initial flow velocities must exceed 2 m/s to lift beachrock fragments out of the country rock, but less energy is needed to shift loosened clasts in subsequent waves (between 1 and 2 m/s). This idea applies equally to beachrock slabs and reef-platform coral boulders. In consequence, interpreting mapped flow velocities derived from hydrodynamic equations for a cyclone event with numerous successive waves should therefore be tackled with caution, since freshly-produced deposits are likely to have experienced changing transport environments through the lifespan of the storm.

Finally, although TC Tomas passed very close to the studied coastline on Taveuni Island, it is evident that its primary contribution was to augment existing boulder fields rather than the creation of new fields in alternative locations. Moreover, ignoring a few isolated exceptions, both fresh and earlier coral boulders are relatively-well confined to discrete sites and do not demonstrate any remarkable linear continuity along the reef edge, in spite of the availability of reef platforms as suitable sites for deposition. This implies that local variability in physiographic attributes of the affected reef and coastline strongly influences both the production and preservation of boulder fields in eastern Taveuni. These attributes may include reef geomorphology and configuration, ecology, roughness, sensitivity to wave attack and relaxation time since the last episode of disturbance.

5.6 Conclusions

TC Tomas was a category-4 cyclone which tracked through Fiji waters in March 2010 and was the most intense storm to strike Taveuni Island in recent years. Investigation of reef-platform boulder fields comprising large numbers of both fresh and older coral boulders enabled discrimination between the storm in question and prior HEMI events. Of interest, although TC Tomas was sufficiently energetic to produce fresh debris (mostly sourced from reef crests or fore-reef slopes rather than the reef surface), this debris constitutes only a minor fraction of original boulder fields (approximately 20 %). In addition, measurement of coral boulder sizes reveals that unrecorded events in the past delivered much coarser material than TC Tomas. This evidence does not necessarily imply that past storms were of higher intensity, as other explanations are possible for the presence of huge boulders. Possibilities include ancient tsunamis, Holocene inundation events at times of higher sea-level stands, more direct approaches of palaeo-cyclones, or rogue waves coinciding with spring tides. Nevertheless, since TC Tomas was unable to remobilise the larger original boulders, we can be reasonably certain that higher-energy conditions (waves and surging currents driven across reef flats) were a feature of the past coastal inundations that originally emplaced these pre-existing deposits.

Recommended future work would be to investigate the timing and frequency of HEMI events throughout the late Holocene at the Taveuni study site, as has been achieved for other coastlines in tropical regimes elsewhere (Hayne and Chappell 2001; Radtke et al. 2003; Zhao et al. 2009). This might be accomplished through accurate dating of boulder carbonate material by uranium-series methods (Gale 2009; Yu et al. 2012). In the Taveuni context, because conservation of the Bouma–Lavena coastline and adjacent fringing reefs is a matter of priority as a Marine Protected Area, information on the magnitude-timing-frequency of prehistorical HEMI events would certainly assist in assessing the future long-term risks of coastal hazards in this area.

References

Ash J (1987) Stunted cloud-forest in Taveuni, Fiji. Pac Sci 41:1–4

BNHP (2011) Bouma National Heritage Park. http://www.boumafiji.com/index.cfm. Accessed Oct 2012

Etienne S (2012) Marine inundation hazards in French Polynesia: geomorphologic impacts of Tropical Cyclone Oli in February 2010. In: Terry JP, Goff J (eds) Natural hazards in the Asia–Pacific region: recent advances and emerging concepts. Geological Society of London, special publication 361:21–39

Etienne S, Paris R (2010) Boulder accumulations related to storms on the south coast of the Reykjanes Peninsula (Iceland). Geomorphology 114:55–70. doi:10.1016/j.geomorph.2009.02.008

Etienne S, Terry JP (2012) Coral blocks, gravel tongues and sand sheets: features of coastal accretion and sediment nourishment by Cyclone Tomas (March 2010) on Taveuni Island, Fiji. Geomorphology 175–176:54–65. doi:10.1016/j.geomorph.2012.06.018

FMS (2010) Fiji Islands Climate Outlook February to April 2010 and May to July 2010. Vol 4: Issue 2. Fiji Meteorological Services, Nadi, Fiji, 6 pp

Gale SJ (2009) Dating the recent past. Quat Geochronol 4:374–377. doi:10.1016/j.quageo.2009.05.011

Hayne M, Chappell J (2001) Cyclone frequency during the last 5000 years at Curacoa Island, north Queensland, Australia. Palaeogeogr Palaeoclimatol Palaeoecol 168:207–219. doi:10.1016/S0031-0182(00)00217-0

Maragos JE, Baines GBK, Beveridge PJ (1973) Tropical Cyclone Bebe creates a new land formation on Funafuti Atoll. Science 181:1161–1164. doi:10.1126/science.181.4105.1161

Nandasena NAK, Paris R, Tanaka N (2011) Reassessment of hydrodynamic equations to initiate boulder transport by high energy events (storms, tsunamis). Mar Geol 281:70–84. doi:10.1016/j.margeo.2011.02.005

Nott J (2003) Waves, coastal boulder deposits and the importance of the pre-transport setting. Earth Planet Sci Lett 210:269–276. doi:10.1016/S0012-821X(03)00104-3

Radtke U, Schellmann G, Scheffers A, Kelletat D, Kromer B, Kasper HU (2003) Electron spin resonance and radiocarbon dating of coral deposited by Holocene tsunami events on Curaçao, Bonaire and Aruba (Netherlands Antilles). Quatern Sci Rev 22:1309–1315. doi:10.1016/S0277-3791(03)00036-2

Terry JP, Etienne S (2010a) Recent heightened tropical cyclone activity east of 180° in the South Pacific. Weather 65:193–195. doi:10.1002/wea.542

Terry JP, Etienne S (2010b) Tempestuous times in the South Pacific Islands. Science 328(5977):428–429. doi:10.1126/science.328.5977.428

Terry JP, McGree S, Raj R (2004) The exceptional floods on Vanua Levu Island, Fiji, during Tropical Cyclone Ami in January 2003. J Nat Disaster Sci 26:27–36

Yu K, Zhao J, Roff G, Lybolt M, Feng Y, Clark T, Li S (2012) High-precision U-series ages of transported coral blocks on Heron Reef (southern Great Barrier Reef) and storm activity during the past century. Palaeogeogr Palaeoclimatol Palaeoecol 337–338:23–36. doi:10.1016/j.palaeo.2012.03.023

Zann L (1991) Effects of Cyclone Ofa on the fisheries and coral reefs of Upolu, Western Samoa in 1990. Unpublished report prepared for the Government of Western Samoa, Food and Agricultural Organization of the United Nations. SAM/89/002 Technical Report no. 2

Zhao J, Neil DT, Feng Y, Yu K, Pandolfi JM (2009) High-precision U-series dating of very young cyclone-transported coral reef blocks from Heron and Wistari reefs, southern Great Barrier Reef, Australia. Quatern Int 195:122–127. doi:10.1016/j.quaint.2008.06.004

Chapter 6
Outlook for Boulder Studies Within Tropical Geomorphology and Coastal Hazard Research

Abstract Reef-platform coral boulders are produced, transported and deposited during high-energy marine inundation events such as large storms or tsunamis. Documented for centuries as extraordinary features of the coastal landscape, these enigmatic boulders have recently proven invaluable indicators for characterising and interpreting marine erosion and transport processes on shorelines. As such, the examination of boulder deposits has become increasingly applicable to coastal hazard and risk assessment studies, although a number of challenges remain unresolved. Future prospects are optimistic for improving boulder analysis, within the broader scope of developing multi-proxy approaches for investigating the impacts of high-magnitude inundation events on coasts.

6.1 Brief Summary: Current Understanding, Guiding Questions

On tropical coastlines, under infrequent conditions of exceptional wave height and surge currents, such as are encountered during intense storms or powerful tsunamis, large fragments of adjacent coral reefs are often detached and emplaced on the reef surface. The resulting reef-platform coral boulders (RPCBs) are normally conspicuous features that are easily identified and, crucially for the coastal geomorphologist, often accessible for investigation. Official records of reef boulders can be traced back to at least the early 1800s and examples may be highlighted where local communities have named boulders as singular coastal landmarks. Since earliest accounts by the British navigator Matthew Flinders in 1814 on the Great Barrier Reef of Australia, a variety of terms have been coined for coral boulders throughout history. Even within the burgeoning scientific interest of recent decades, terminology in use has remained colourful but become somewhat inconsistent. Unfortunately this situation causes confusion and raises hurdles against easy comparison of coastal boulder observations from place to place. For the purposes of clarification, our review has identified the plethora of expressions that have been used to refer to coastal boulders at various times, but we reiterate the advice that future research should abide by the recognised Blair and McPherson (1999) grain-size scheme for describing the sedimentology of large clastic deposits (see Table 2.2).

A number of applications of coastal boulder research, especially studies of RPCBs, for understanding fundamental aspects of the behaviour of high-energy marine inundation (HEMI) events have been explained. In particular, by measuring the position, distribution and dimensions of RPCBs, information on HEMI wave height and inundation direction at specific coastal locations can be derived. By knowing the volume, shape and weight of boulders, as well as their original environmental setting, the minimum flow velocity required for initialising their movement can be inferred from several hydrodynamic transport equations developed and refined by other workers. Together with age-dating and frequency analysis of past (unrecorded) events, this is helpful in determining the vulnerability of coastal locations, which in turn is necessary for carrying out risk assessments and mitigating against the possible impacts of future catastrophic inundation hazards.

Notwithstanding this enormous potential, however, a number of problems hampering coastal boulder studies have been identified. The following list is not intended to provide an exhaustive coverage, but draws attention to the major difficulties faced. Next, this logically encourages the formulation of relevant research questions, which are summarised here (in brackets) and expanded upon in the next section.

- There is inconsistent presentation of boulder data in the existing scientific literature, resulting in incomparability between available datasets. (How can this situation be rectified?)
- Inaccuracies with both measuring boulder dimensions and calculating volume tends to over-estimate boulder size. (To what extent do emerging photogrammetric techniques reduce boulder measurement errors?)
- Identifying original boulder source locations is problematic, potentially leading to misrepresentation of the strength of HEMI events. (What possibilities exist for improving on boulder source identification?)
- Assumptions of boulder emplacement by a single inundation event are often invalid. Reworking of boulders by backwash or subsequent events leads to erroneous estimations of wave energy. (Boulder reworking mechanisms need to be much better understood; what evidence should be targeted?)
- The size of coastal boulders, especially carbonate clasts, decreases over time through various processes of degradation. Old boulders are probably appreciably smaller now than at the time of their production and deposition. Again, this has implications for interpreting the nature of palaeo-events correctly. (Is it possible to establish rates of boulder attrition, weathering and bioerosion according to environmental setting?)
- Deliberate human interference on coastlines means that some large boulders are removed, and so the opportunity for their analysis is lost.
- Hydrodynamic transport equations for estimating the wave energy needed to initiate boulder movement are of tremendous value. Inevitably though, approximation means that existing equations simplify the processes involved and

essential parameters have been neglected. (How influential is seawater turbidity for sediment transportation? Does mobilisation of smaller sediment-size fractions (cobbles, gravels and sands) contribute to boulder transport through buoyant support and inter-clast collision? To what extent does bed roughness limit incipient boulder movement and overall transport distance?).

- Storms and tsunamis are two types of HEMI events with very different wave return periods and behaviour. Yet, until now agreement has not been reached on how to distinguish with certainty between boulders delivered by these contrasting types of coastal hazards. (Which new approaches show most promise for differentiating tsunami from storm boulders? How far does the study of constructional landforms on coasts (e.g. boulder ridges) assist in understanding the respective dynamics of storm or tsunami waves?).

6.2   Future Prospects and Recommendations

Recognising existing shortcomings is not a criticism of earlier work, but is the cornerstone of sound deductive reasoning. Only through the identification of gaps in current knowledge can scientific methodology then be applied to tackle the issues that need to be addressed. In this way real progress can be made. Thus, grasping the nettle of the difficulties mentioned above is necessary within the broader context of uncertainties in coastal boulder studies. A sensible starting point is to prioritise a set of tasks that can provide avenues for future research. The special emphasis here is on RPCBs observed on tropical (coral reef) coastlines, but the majority of recommendations also have direct relevance for coarse clastic deposits seen on coastlines beyond tropical regimes.

Improving hydrodynamic transport modelling is a priority concern. This may be accomplished in a variety of ways. One suggestion is for greater flexibility in existing transport equations through the substitution of a range of values for the water density parameter to represent turbidity. Turbid waters, i.e. marine water mixed with variable sediment loads, are a common characteristic of tsunami waves. The sediment load modifies the water viscosity, the fluid behaviour and the buoyancy afforded to boulders by the water column. Accordingly, the assumption that a tsunami acts as a Newtonian fluid is not entirely valid. Indeed, Kain et al. (2012) recently hypothesised that certain tsunamis act more as a Bingham fluid, especially where depositional evidence indicates *en masse* (or debris flow) type transport of sediments. Boulder movement is also affected by inter-clast collision. For example, where boulders form part of wider accumulations (boulder fields), constructional features (gravel ramparts or ridges) or man-made sea defences (rock armour), numerous clasts may be reworked simultaneously during an energetic marine inundation event. When an individual mobile boulder collides with another static boulder lying on the reef flat, sufficient momentum may be transferred to initialise movement of the static clast, which might otherwise not occur by fluid pressure alone. So far, however, such influences have largely been ignored. In addition, the topographic rugosity of the surface over which the boulders are mobilised (e.g. coral reef, inter-tidal flat, wave-cut shore platform, emerged terrace) undoubtedly influences boulder transport. It is therefore advisable that a rugosity coefficient be developed, ranging from 0 (flat rock surface) to 1 (highly irregular surface with boulder-sized traps) (Fig. 6.1).

In light of these considerations, it is apparent that available hydrodynamic transport equations would benefit from further refinement, to incorporate both the function of suspended and bedload sediments, and the factor of surface roughness. Yet tackling these complex issues will not be a trivial undertaking. Consequently, it is likely that a combination of theoretical approaches, modelling simulations and physical experiments carried out in wave laboratories (Fig. 6.2) will all be required. Although perhaps daunting, therein lie an exciting set of new endeavours for modern coastal geomorphology.

It is similarly anticipated that future research will address and eventually overcome the other questions posed in the preceding section. In this regard, it would be advantageous for boulder studies to leverage on the recent advancements made in relevant arenas such as computer-aided interpretation of satellite images (especially new spectral bands), high-precision dating techniques and close-range photogrammetry. Clearly there are inherent advantages and limitations with studying both coarse and fine textured extreme-wave deposits. While coastal boulders are normally better preserved than sandsheets in high-energy coastal settings, their absence need not necessarily imply an absence of hazard risk entirely (Kelletat et al. 2007). Moreover, new methods for the analysis of well-preserved sand and mud deposits are continually achieving greater degrees of precision, particularly as applied to modelling wave dynamics and sediment transport capacity. Innovative methods include anisotropy of magnetic susceptibility (AMS) (Wassmer and Gomez 2011), inverse sediment-transport modelling (Jaffe et al. 2011) and analysis of nannoliths (heterogeneous suites of biogenic carbonate particles with silt–clay size dimensions; Paris et al. 2010). Overall it is understood that a multi-proxy approach is deemed preferable for obtaining a more robust and comprehensive record of past HEMI events (Yu et al. 2009; Etienne et al. 2011). Multi-proxy approaches rely not only on the geological proxies of fine-grained sediment layers and coastal boulders, but also encompass geomorphological, geochemical, ecological, historical and archaeological evidence, as elaborated adeptly by Goff et al. (2010).

In conclusion, it is hoped that this treatise has laid out a convincing argument: studies of reef-platform coral boulders (and other carbonate/non-carbonate boulders) form an integral component of scientific research that aims to better comprehend the nature of high-energy marine inundation events on tropical coastlines. Although much research in the past decade has focused on boulders deposited by known HEMI events, such as the 2004 Indian Ocean tsunami in Indonesia and Thailand and the 2009 South Pacific tsunami in Samoa and Tonga, a positive outcome in the aftermath of these disasters has been the identification of a range of key features of boulder distribution and transportation. The opportunity now exists to broaden investigations to further locations where coastal boulder deposits have been generated by hitherto unknown ancient events, or by recognised

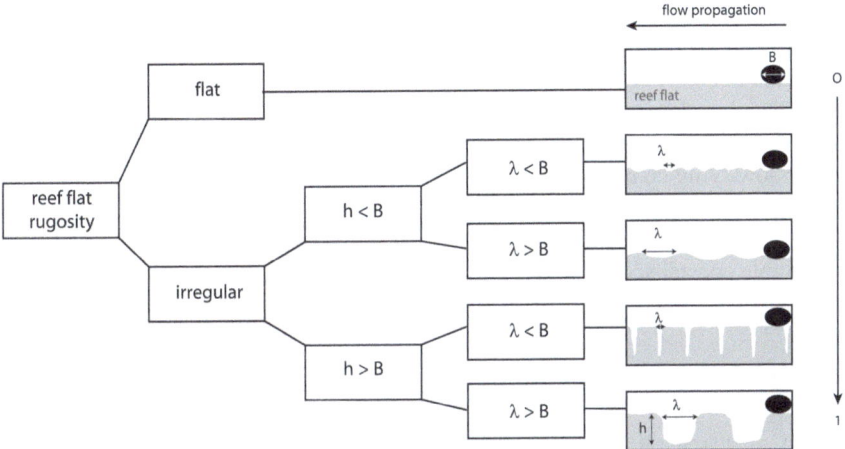

Fig. 6.1 Conceptual model of a rugosity coefficient for a coral reef platform, with increasing rugosity affecting the transport distance of a coastal boulder. Zero (0) implies that topographic rugosity is nill and does not interfere with boulder propagation over the reef flat. One (1) would be the case of a very irregular surface where depressions (grooves, potholes) might entirely trap the boulder. In this situation, varying levels of energy are necessary to displace a boulder from the reef crest to the beach. For example, sliding is possible across the spurs, but lifting would be necessary if a boulder were initially trapped in a depression. The idea can be extended to include other kinds of coastal platforms, such as raised or karstified marine terraces, wave-cut rock shorelines or intertidal mud flats

Fig. 6.2 The 36 m-long wave tank at Nanyang Technological University (NTU) in Singapore. The facility is designed for laboratory studies of wave behaviour and wave interaction with coastlines of various configurations (Photo courtesy of Mr. Shawn Sim, NTU Department of Civil Engineering and the Earth Observatory of Singapore)

historical events that have so far escaped close attention. Across the vast expanse of the Asia–Pacific region in particular, innumerable islands offer prized sites for exploration. Many are fringed by coral reefs where carbonate boulders are common, but the character of their coastal deposits is still underrepresented in the geomorphic literature. At the same time, an increasing awareness for people living on low-lying coastlines is necessary to fully appreciate the potential for future marine inundations. Continuing work on coastal boulders will undoubtedly enrich the information available and thereby assist in the long-term ambition of adapting vulnerable coastal societies to the natural hazards they face.

References

Blair TC, McPherson JG (1999) Grain-size and textural classification of coarse sedimentary particles. J Sediment Res 69:6–19. doi:10.1306/D426894B-2B26-11D7-8648000102C1865D

Etienne S, Buckley M, Paris R, Nandasena NAK, Clark K, Chagué-Goff C, Goff J, Richmond B (2011) The use of boulders for characterizing past tsunamis: lessons from the 2004 Indian Ocean and 2009 South Pacific tsunamis. Earth Sci Rev 107:75–90. doi:10.1016/j.earscirev.2010.12.006

Goff J, Pearce S, Nichol SL, Chagué-Goff C, Horrocks M, Strotz L (2010) Multi-proxy records of regionally-sourced tsunamis, New Zealand. Geomorphology 118:369–382. doi:10.1016/j.geomorph.2010.02.005

Jaffe B, Buckley M, Richmond B, Strotz L, Etienne S, Clark K, Watt S, Gelfenbaum G, Goff J (2011) Flow speed estimated by inverse modeling of sandy sediment deposited by the 29 September 2009 tsunami near Satitoa, east Upolu, Samoa. Earth Sci Rev 107:23–37. doi:10.1016/j.earscirev.2011.03.009

Kain CL, Gomez C, Moghaddam AE (2012) Comment on 'Reassessment of hydrodynamic equations: minimum flow velocity to initiate boulder transport by high energy events (storms, tsunamis)'. In: Nandasena NAK, Paris R, Tanaka N (eds) [Mar Geol 281:70–84] Mar Geol, 319–322:75–76. doi:10.1016/j.margeo.2011.08.008

Kelletat D, Scheffers SR, Scheffers A (2007) Field signatures of the SE-Asian mega-tsunami along the west coast of Thailand compared to Holocene paleo-tsunami from the Atlantic region. Pure Appl Geophys 164(2–3):413–431. doi:10.1007/s00024-006-0171-6

Paris R, Cachão M, Fournier J, Voldoire O (2010) Nannoliths abundance and distribution in tsunami deposits: example from the December 26, 2004 tsunami in Lhok Nga (northwest Sumatra, Indonesia). Géomorphologie: relief, processus, environnement 1:109–118. doi:10.4000/geomorphologie.7865

Wassmer P, Gomez C (2011) Development of the AMS method for unconsolidated sediments. Application to tsunami deposits. Géomorphologie: relief, processus, environnement 3:279–290. doi:10.4000/geomorphologie.9491

Yu K, Zhao J, Shi Q, Meng Q (2009) Reconstruction of storm/tsunami records over the last 4000 years using transported coral blocks and lagoon sediments in the southern South China Sea. Quat Int 195(1–2):128–137. doi:10.1016/j.quaint.2008.05.004

Index

J. P. Terry et al., *Reef-Platform Coral Boulders*, SpringerBriefs in Earth Sciences,
DOI: 10.1007/978-981-4451-33-8, © The Author(s) 2013